Numerical Simulation of the Groundwater-Flow System in Chimacum Creek Basin and Vicinity, Jefferson County, Washington

By Joseph L. Jones, Kenneth H. Johnson, and Lonna M. Frans

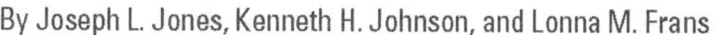

Prepared in cooperation with Jefferson County and the
Washington State Department of Ecology

Scientific Investigations Report 2013–5160

U.S. Department of the Interior
U.S. Geological Survey

U.S. Department of the Interior
SALLY JEWELL, Secretary

U.S. Geological Survey
Suzette M. Kimball, Acting Director

U.S. Geological Survey, Reston, Virginia: 2013

For more information on the USGS—the Federal source for science about the Earth, its natural and living resources, natural hazards, and the environment, visit http://www.usgs.gov or call 1–888–ASK–USGS.

For an overview of USGS information products, including maps, imagery, and publications, visit http://www.usgs.gov/pubprod

To order this and other USGS information products, visit http://store.usgs.gov

Suggested citation:
Jones, J.L., Johnson, K.H., and Frans, L.M., 2013, Numerical simulation of the groundwater-flow system in Chimacum Creek Basin and vicinity, Jefferson County, Washington: U.S. Geological Survey Scientific Investigations Report 2013–5160, 80 p., http://pubs.usgs.gov/sir/2013/5160/.

Contents

Abstract 1
Introduction 2
    Purpose and Scope 2
    Description of Study Area 2
Groundwater-Flow System 4
    Geologic Setting 4
    Hydrogeologic Units 4
    Hydraulic Conductivity 6
    Groundwater-Flow Directions 8
    Groundwater Discharge 8
    Groundwater and Surface-Water Interactions 9
    Groundwater-Level Fluctuations 9
    Groundwater Budget 9
Numerical Simulation of Groundwater-Flow System 9
    Model Grid and Layering 10
    Time Discretization 10
    Boundary Conditions 10
    Recharge 18
    Groundwater Withdrawals 18
    Streams 20
    Model Calibration 21
    Sensitivity Analysis and Final Parameter Values 21
    Assessment of Steady State Calibration 39
    Model Limitations 45
Model Applications 46
    Model-Derived Groundwater Budget 46
    Description and Analyses of Model Simulations 47
        Description of Simulations 47
        Current Conditions 47
        Probable Future Use 47
        Full Beneficial Use 49
        Sanitary Sewer 49
    Comparison of Current Conditions, Probable Future Use, Full Beneficial Use, and
        Sanitary Sewer Simulations 49
    Flow Directions from Sources and to Sinks 70
    Areal Variation of Response Coefficient for a Well 74
    Effect of Irrigation Wells and Depth on Chimacum Creek Streamflow 74
Suggestions for Further Study 76
Summary 76
Acknowledgments 77
References Cited 77

Figures

1. Map showing locations of Washington State Department of Ecology (WDOE) gaging station 17B050 and monitoring wells, Chimacum Creek Basin and vicinity, Jefferson County, Washington 3

2. Example cross-section showing hydrogeologic units, Chimacum Creek Basin and vicinity, Jefferson County, Washington..6

3. Map showing location of model boundary condition cells, Chimacum Creek Basin and vicinity, Jefferson County, Washington...11

4. Maps showing extent and thickness of model layers, Chimacum Creek Basin and vicinity, Jefferson County, Washington...12

5. Map showing average annual recharge from precipitation in the Chimacum Creek Basin and vicinity, Jefferson County, Washington, water years 1995–2009.....................19

6. Graph showing historical groundwater use, probable future groundwater use, steady-state full beneficial groundwater use and total groundwater use for Public Utility District #1, Chimacum Creek Basin and vicinity, Jefferson County, Washington, water years 1995–2030 20

7. Map showing locations of wells and streamflow sites used for calibration, Chimacum Creek Basin and vicinity, Jefferson County, Washington 22

8. Maps showing aerial distribution of horizontal (*A-F*) and vertical conductivities (*G-L*) and locations of hydraulic conductivity calibration pilot points for all model layers, Chimacum Creek Basin and vicinity, Jefferson County, Washington 23

9. Identifiability plots for parameters used in model calibration, Chimacum Creek Basin and vicinity, Jefferson County, Washington...37

10. Graph showing simulated and measured groundwater-level altitude for the calibrated model for steady-state conditions, Chimacum Creek Basin and vicinity, Jefferson County, Washington 40

11. Maps showing simulated water-level altitudes and residuals for the calibrated model for steady-state conditions, model layers (*A*) 2 (unit UA), (*B*) 4 (unit LA), and (*C*) 5 (unit LC), Chimacum Creek Basin and vicinity, Jefferson County, Washington 41

12. Graph showing simulated steady-state base flows and measured base flows, Chimacum Creek Basin and vicinity, Jefferson County, Washington 44

13. Maps showing simulated water-level altitude change in each hydrogeologic unit between the Current Conditions and Probable Future Use simulations, Chimacum Creek Basin and vicinity, Jefferson County, Washington 51

14. Maps showing simulated water-level altitude change in each hydrogeologic unit between Probable Future Use and Full Beneficial Use simulations, Chimacum Creek Basin and vicinity, Jefferson County, Washington 57

15. Maps showing simulated water-level altitude change in each hydrogeologic unit between Probable Future Use and Sanitary Sewer Simulations, Chimacum Creek Basin and vicinity, Jefferson County, Washington..63

16. Map showing forward particle tracking from topmost layer to sinks, colored by boundary condition at particle terminus, Chimacum Creek Basin and vicinity, Jefferson County, Washington 71

Figures—Continued

17. Map showing reverse particle tracking from the Jefferson County Public Utility District Number 1 Sparling well, Chimacum Creek Basin and vicinity, Jefferson County, Washington 72

18. Map showing forward and reverse particle tracking to and from Chimacum Creek and East Fork Chimacum Creek, Jefferson County, Washington 73

19. Map showing response coefficient for a well pumping 68,583 cubic feet per day at any location in model layer 4 (Lower Aquifer), Chimacum Creek Basin and vicinity, Jefferson County, Washington 75

Tables

1. Model layer correlation with hydrogeologic units, lithology, occurrence, and thickness, Chimacum Creek Basin and vicinity, Jefferson County, Washington 5

2. Water use in the Jefferson County Public Utility District #1 (public-supply use) and outside the public-supply area (self-supplied use) and recharge of groundwater by return flow from each class of user, during each year of the recorded period 1994–2009, with projections through 2030, Chimacum Creek Basin and vicinity, Jefferson County, Washington 7

3. Total estimable groundwater inflows (recharge) and outflows (discharge), and residual, Chimacum Creek Basin and vicinity, Jefferson County, Washington 8

4. Water levels used for steady-state model calibration, Chimacum Creek Basin and vicinity, Jefferson County, Washington 35

5. Surface-water discharge measurements used for steady-state model calibration, Chimacum Creek Basin and vicinity, Jefferson County, Washington 36

6. Final values for model calibration parameters, Chimacum Creek Basin and vicinity, Jefferson County, Washington 38

7. Calibration statistics by hydrogeologic unit and base flow, Chimacum Creek Basin and vicinity, Jefferson County, Washington 39

8. Water-budget components for the terrestrial part of the calibrated steady-state model, Chimacum Creek Basin and vicinity, Jefferson County, Washington, 2001–09 47

9. Model-derived groundwater flow and comparable estimates, Chimacum Creek Basin and vicinity, Jefferson County, Washington 48

10. Ranges of water-level changes beween simulations representing current and future conditions, Chimacum Creek Basin and vicinity, Jefferson County, Washington 49

11. Comparison of selected water budget components for the Current Conditions and Probable Future Use simulations for Chimacum Creek model subbasin, Jefferson County, Washington 50

12. Comparison of selected water budget components for the Probable Future Use, Full Beneficial Use, and Sanitary Sewer simulations, Chimacum Creek model subbasin, Jefferson County, Washington 69

Conversion Factors and Datums

Conversion Factors

Multiply	By	To obtain
Length		
foot (ft)	0.3048	meter (m)
mile (mi)	1.609	kilometer (km)
Area		
acre-foot	0.4047	cubic meter (m^3)
square foot (ft^2)	0.09290	square meter (m^2)
square mile (mi^2)	259.0	hectare (ha)
Volume		
acre-foot (acre-ft)	1,233	cubic meter (m^3)
Flow rate		
acre-foot per year (acre-ft/yr)	1,233	cubic meter per year (m^3/yr)
foot per day (ft/d)	0.3048	meter per day (m/d)
foot per year (ft/yr)	0.3048	meter per year (m/yr)
cubic foot per second (ft^3/s)	0.02832	cubic meter per second (m^3/s)
inch per year (in/yr)	25.4	millimeter per year (mm/yr)
Transmissivity*		
foot squared per day (ft^2/d)	0.09290	meter squared per day (m^2/d)

*Transmissivity: The standard unit for transmissivity is cubic foot per day per square foot times foot of aquifer thickness [(ft^3/d)/ft^2]ft. In this report, the mathematically reduced form, foot squared per day (ft^2/d), is used for convenience.

Datums

Vertical coordinate information is referenced to the North American Vertical Datum of 1988 (NAVD 88).

Horizontal coordinate information is referenced to the North American Datum of 1983 (NAD 83).

Altitude, as used in this report, refers to distance above the vertical datum.

Well-Numbering System

Wells in Washington State are assigned a local well number that identifies each well based on its location within a township (T), range (R), section, and 40-acre tract. For example, well 29N/01W-35J01 refers to township (T. 29 N) and the range (R. 01 W) north of the Willamette Base Line and west of the Willamette Meridian. The first number following the hyphen indicates the section (35) within the township, and the letter (J) following the section number indicates the 40-acre subdivision of the section. The final two-digit number (01) uniquely distinguishes individual wells in the same 40-acre tract.

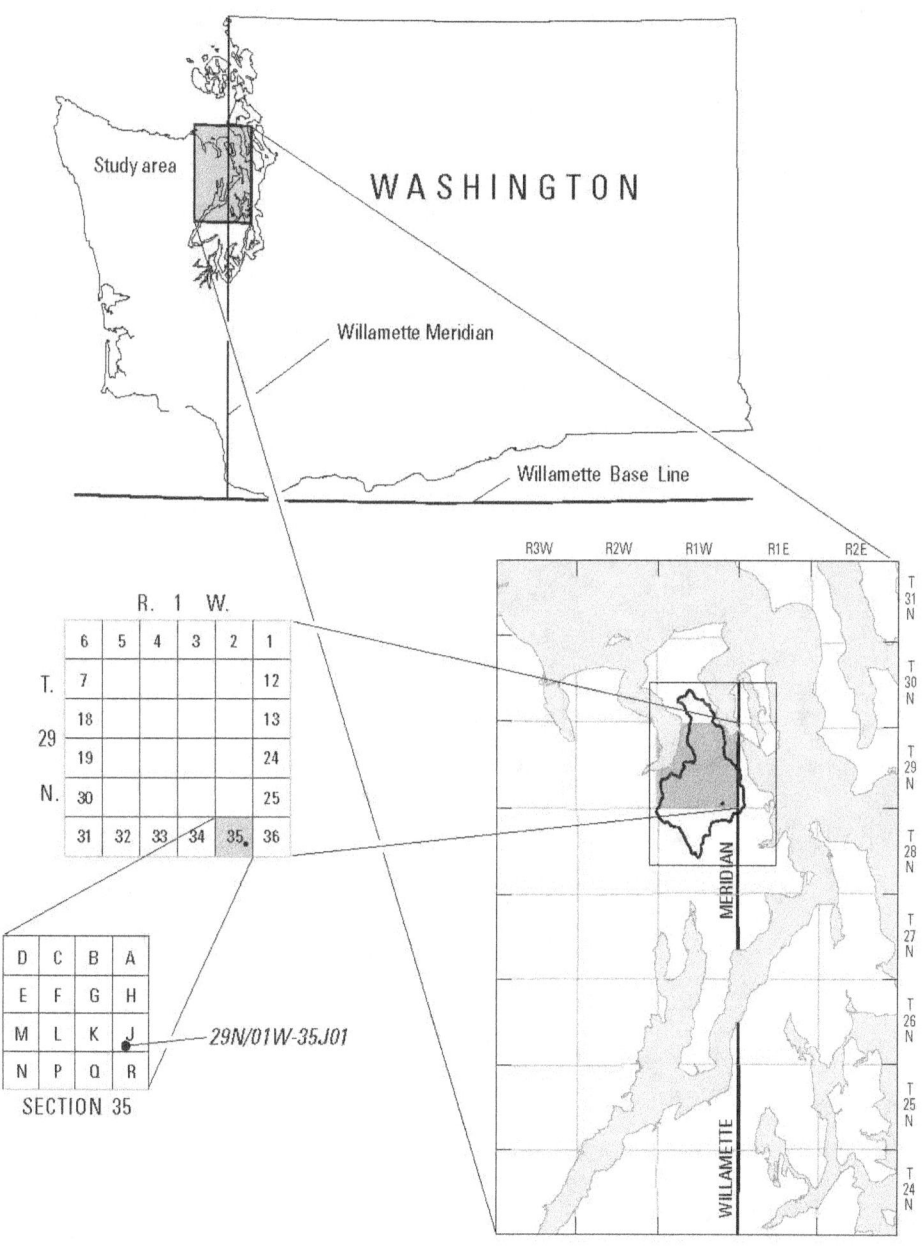

Numerical Simulation of the Groundwater-Flow System in Chimacum Creek Basin and Vicinity, Jefferson County, Washington

By Joseph L. Jones, Kenneth H. Johnson, and Lonna M. Frans

Abstract

A groundwater-flow model was developed to evaluate potential future effects of growth and of water-management strategies on water resources in the Chimacum Creek Basin. The model covers an area of about 64 square miles (mi²) on the Olympic Peninsula in northeastern Jefferson County, Washington. The Chimacum Creek Basin drains an area of about 53 mi² and consists of Chimacum Creek and its tributary East Fork Chimacum Creek, which converge near the town of Chimacum and discharge to Port Townsend Bay near the town of Irondale. The topography of the model area consists of north-south oriented, narrow, regularly spaced parallel ridges and valleys that are characteristic of fluted glaciated surfaces. Thick accumulations of peat occur along the axis of East Fork Chimacum Creek and provide rich soils for agricultural use. The study area is underlain by a north-thickening sequence of unconsolidated glacial (till and outwash) and interglacial (fluvial and lacustrine) deposits, and sedimentary and igneous bedrock units that crop out along the margins and the western interior of the model area. Six hydrogeologic units in the model area form the basis of the groundwater-flow model. They are represented by model layers UC (upper confining), UA (upper aquifer), MC (middle confining), LA (lower aquifer), LC (lower confining), and OE (bedrock).

Groundwater flow in the Chimacum Creek Basin and vicinity was simulated using the groundwater-flow model, MODFLOW-2005. The finite-difference model grid comprises 245 columns, 313 rows, and 6 layers. Each model cell has a horizontal dimension of 200 × 200 feet (ft). The thickness of model layers varies throughout the model area and ranges from 5 ft in the non-bedrock units to more than 2,400 ft in the bedrock. Groundwater flow was simulated for steady-state conditions, which were simulated for calibration of the model using average recharge, discharge, and water levels for the 180-month period October 1994–September 2009. The model as calibrated has a mean residual of 4.5 ft and a standard error on the mean of 2.1 ft for heads, and 0.64 ± 0.42 cubic feet per second for streamflows. After the model was calibrated, a Current Conditions simulation was developed to reflect current (October 2008–September 2009) hydrologic conditions, with representative pumping, return flows, and "normal" recharge (based on National Weather Service average precipitation for 1981 to 2010). The Current Conditions simulation was used to estimate current flow quantities, and as a basis to compare other simulations.

Simulated steady-state inflow to the model area from precipitation and secondary recharge, or "return flow," was 16,347 acre-feet per year (acre-ft/yr); groundwater inflow from other basins to the north of the model boundary was 1,518 acre-ft/yr (net, 3,114 acre-ft/yr in and 1,596 acre-ft/yr out) and simulated inflow from lake leakage was 613 acre-ft/yr (net, 684 acre-ft/yr in and 71 acre-ft/yr out). Simulated outflow from the model primarily was through discharge to Puget Sound (10,022 acre-ft/yr), streams (5,424 acre-ft/yr), springs and seeps (1,521 acre-ft/yr), and through withdrawals from wells (1,506 acre-ft/yr).

Four simulations were formulated using the calibrated model—one to represent current conditions (2009, the end of the period used for calibration) and three to provide representative examples of how the model can be used to evaluate the relative effects of potential changes in groundwater withdrawals and consumptive use on groundwater levels and stream base flows: Probable Future Use, based on population projections; Full Beneficial Use, based on Jefferson County Public Utility District #1 water rights; Sanitary Sewer, based on eliminating septic return flows in the Urban Growth Area. Particle tracking was used to assess flowpaths from sources and to sinks, and the effects of the presence of irrigation wells and their depths was assessed.

Introduction

Projected increases in population and development in northeastern Jefferson County, Washington, are expected to lead to increased groundwater withdrawals in the Chimacum Creek Basin. Additionally, changes in land use could reduce groundwater recharge in the basin, thereby reducing groundwater levels and discharge from the groundwater system to Chimacum Creek. Groundwater discharge to the creek, also referred to as base flow, is critical for maintaining ecological health in the creek throughout the year. Groundwater discharge is particularly important during summer and early autumn, when it supplies most, if not all, streamflow. Chimacum Creek provides habitat for salmonids, including species listed under the Endangered Species Act (ESA), such as summer-run chum salmon (*Oncorhynchus keta*; threatened), coho salmon (*Oncorhynchus kisutch*; species of concern), and steelhead (*Oncorhynchus mykiss*; proposed for listing as threatened in March 2006) (National Oceanic and Atmospheric Administration, 2006). Decision makers and water-resources managers can use quantitative tools to assess the effect of different water-management options so that they can plan for future growth and development in ways that minimize adverse effects on Chimacum Creek.

In April 2007, the U.S. Geological Survey (USGS) Washington Water Science Center, in cooperation with Jefferson County and the Washington State Department of Ecology (WDOE), began a study to understand the potential effect of different patterns of growth and water-management strategies on the groundwater and surface-water resources of the Chimacum Creek Basin. This study is based on information from previous studies, as well as newly collected data.

Purpose and Scope

This report describes the construction of a numerical groundwater-flow model and its application to the evaluation of potential future effects of growth and of water-management strategies on water resources in the Chimacum Creek Basin. It describes the conceptual model that forms the basis of the numerical model—the study area, the geologic and hydrologic settings, the hydrogeologic units that are the basis of the model layers, and recharge to, and discharge from, the groundwater system. In this report, areal and temporal discretization, boundary conditions, recharge estimates, withdrawal estimates, and calibration of the numerical model

are described. Finally, the application of the model to different future simulations is described. These simulations include application of full water rights owned by the Jefferson County Public Utility District #1 (PUD #1), the effects of projected population growth, the effects of replacing septic systems in the Urban Growth Area (UGA) with a sewer system, where pumping affects (or does not affect) streamflow (including depth of pumpage), and for particle tracking analysis of sources to PUD #1 wells, sources, and sinks for Chimacum Creek, and sinks for the model area generally. The possibilities for further study are also discussed.

Description of Study Area

The model covers an area of about 64 square miles (mi^2) on the Olympic Peninsula in northeastern Jefferson County, Washington (fig. 1). The Chimacum Creek Basin drains an area of about 53 mi^2 and consists of Chimacum Creek and its tributary East Fork Chimacum Creek. These creeks converge near the town of Chimacum and discharge to Port Townsend Bay near the town of Irondale. The topography of the study area consists of narrow, regularly spaced parallel ridges and grooves that are characteristic of fluted glaciated surfaces; they are oriented in a north-south direction (Ritter, 1978). This surface has been incised locally by fluvial and postglacial erosion, producing steep sides and hummocky bottoms for the valley. Thick accumulations of peat occur along the axis of East Fork Chimacum Creek and provide rich, agriculturally productive soils. The study area is underlain by a north-thickening sequence of unconsolidated glacial and interglacial deposits. Sedimentary and igneous bedrock units underlie the unconsolidated deposits and crop out along the margins and the western interior of the study area.

The study area has a temperate marine climate with warm, dry summers, and cool, wet winters. Chimacum Creek Basin lies within the rain shadow of the Olympic Mountains, and the annual average precipitation during 1981–2010 at the community of Center (fig. 1) was 28.78 in/yr (National Oceanic and Atmospheric Administration, 2007). In 1996, the population of the Chimacum Creek Basin was 5,675 people, and is projected to increase by almost 30 percent by 2016 (Parametrix and others, 2000). Population density in the basin is highest near the mouth of Chimacum Creek, in the general area of Irondale, Port Hadlock, and Chimacum (fig. 1), which is roughly coincident with the extent of the topmost aquifer Upper Aquifer (UA, model layer 3).

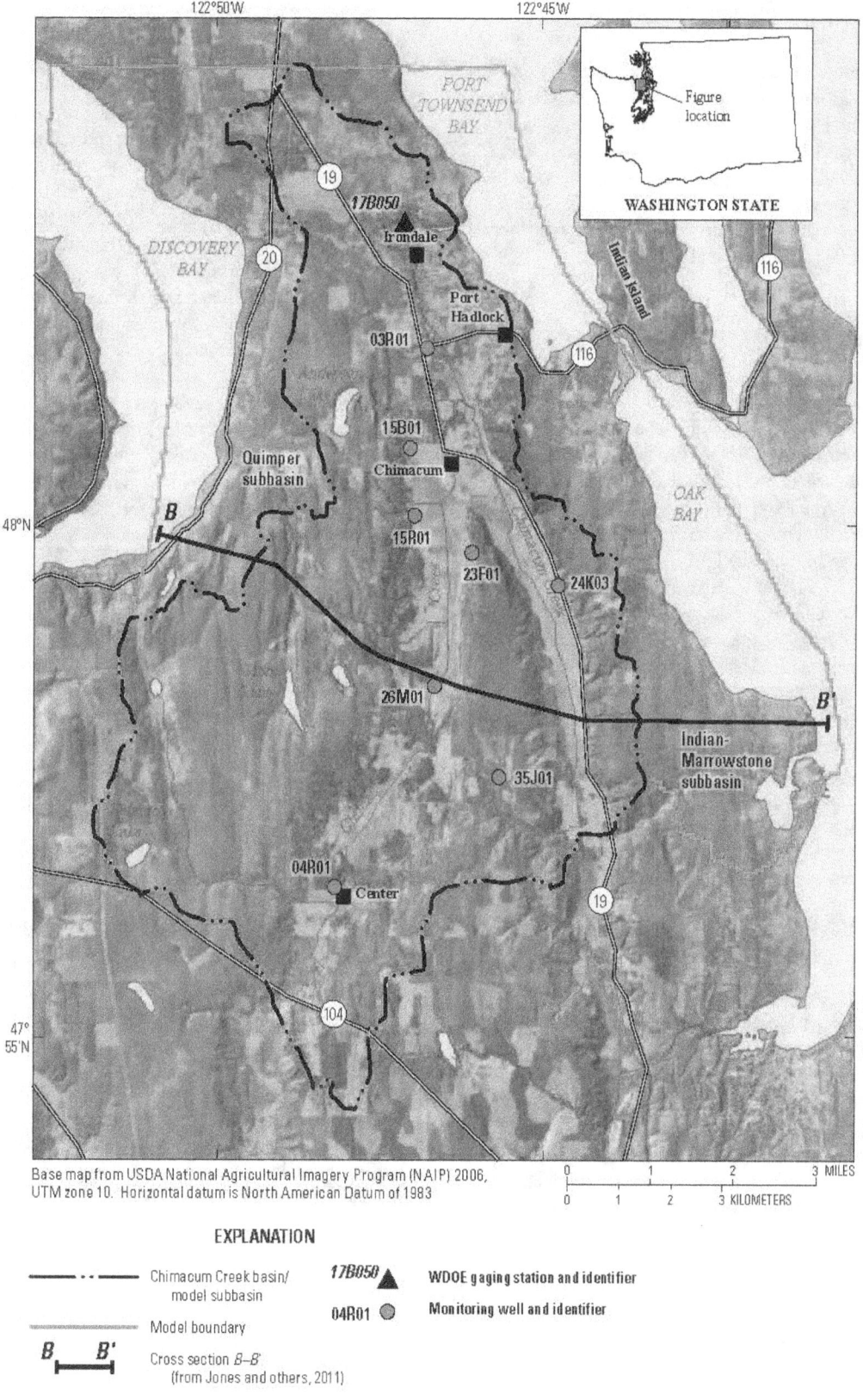

Figure 1. Locations of Washington State Department of Ecology (WDOE) gaging station 17B050 and monitoring wells, Chimacum Creek Basin and vicinity, Jefferson County, Washington.

Groundwater-Flow System

This section describes the hydrogeologic units that constitute the groundwater-flow system in the model area and includes discussions of recharge, flow direction, discharge, exchange of water between the aquifer system and creeks, temporal fluctuations in groundwater levels, and water budget. This information was used to construct and calibrate the numerical model and was drawn from the work of Simonds and others (2004), and Jones and others (2011).

Geologic Setting

The following brief summary of major geologic events in the study area is based on the work of Easterbrook (1979), Grimstad and Carson (1981), Tabor and Cady (1978), and Simonds and others (2004). Tectonic forces related to the subduction of oceanic crust beneath the western coast of North America resulted in uplift and accretion of Eocene to Oligocene sedimentary and igneous rocks along the continental margin. These deformed rocks form the bedrock beneath the study area in eastern Jefferson County. The Puget Lobe of the Cordilleran ice sheet advanced into the study area several times during the Pleistocene Epoch. The most recent period of glaciation, the Vashon Stade of Fraser glaciation, began about 17,000 years ago when the continental ice sheet in Canada expanded, and the Puget Lobe advanced southward, eventually covering the entire Puget Sound Basin before halting and retreating. Unconsolidated deposits of glacial and interglacial origin are present throughout the study area. A typical glacial sequence progresses from advance outwash, to till, to recessional outwash. Fluvial, lacustrine, bog, and marsh depositional environments were common during interglacial periods. The modern-day drainage pattern of Chimacum Creek is mostly determined by pre-existing drainage pathways established by Vashon recessional outwash channels.

Hydrogeologic Units

Jones and others (2011) delineated six hydrogeologic units in the model area (table 1). Geologic units were grouped into hydrogeologic units, comprising aquifers and confining units, on the basis of lithologic (depositional facies, grain size, and sorting) and hydrologic (hydraulic conductivity and unit geometry) characteristics. The hydrogeologic units are represented by model layers that conform to the requirements of MODFLOW (Harbaugh, 2005) and its attendant hydraulic property modules (in this case, Layer Property Flow). Glacial deposits generally are heterogeneous, and although a glacial aquifer may be composed primarily of sand or gravel, it may locally contain varying amounts of clay or silt. Similarly, a confining layer composed predominantly of silt or clay may contain local lenses of coarse material. These variations in lithology may influence the occurrence and movement of groundwater at a scale that is likely too small to be adequately represented by the regional-scale groundwater-flow model constructed for this study. Local-scale variability in the distribution of glacial-depositional facies often results in the formation of spatially discontinuous units of varying thickness. Therefore, some units are not spatially contiguous, and unit thickness may vary considerably over short distances throughout the model area (Jones and others, 2011, figs. 2–6).

In the study area, aquifers consist primarily of coarse-grained glacial outwash, but they also may include coarse-grained sediments within glacial till and coarse-grained interglacial deposits. The hydrogeologic units representing aquifers are Upper Aquifer (UA) and Lower Aquifer (LA). The hydrogeologic units roughly correspond with geological units, recessional outwash (UA), and advance outwash (LA) of the Vashon glacial deposits. The Lower Confining (LC) unit is a productive aquifer in some places. In other places, the LC is a confining unit composed of hundreds of feet of clay. Because most wells are finished in UA and LA, there were insufficient data to credibly subdivide the LC interglacial deposits (and possibly deposits from pre-Vashon glacial epochs) into distinct geologic or hydrogeologic units. Figure 2 (excerpted from Jones and others, 2011, pl. 1) is a representative cross section showing the hydrogeologic units. Table 1 shows the correspondence between model layers, hydrogeologic units, lithology, and range of thickness for the model layers.

The confining units consist primarily of fine-grained glacial outwash, unsorted and compacted glacial till, glaciolacustrine deposits, and fine-grained interglacial deposits. The hydrogeologic units representing confining layers are Upper Confining (UC; model layer 1) and Middle Confining (MC; model layer 3). Lower Confining (LC; model layer 5) is not distinctly a confining unit. UC roughly corresponds with geologic unit Quaternary alluvial, and MC corresponds with Vashon till.

Unconsolidated aquifer and confining units are underlain by low-permeability Eocene to Oligocene sedimentary and igneous bedrock (OE; model layer 6).

Table 1. Model layer correlation with hydrogeologic units, lithology, occurrence, and thickness, Chimacum Creek Basin and vicinity, Jefferson County, Washington.

[**Hydrogeologic unit**: From Jones and others (2011)]

Model layer	Hydrogeologic unit	Lithology	Occurrence and thickness
1	UC–Upper Confining unit (alluvial and recessional outwash deposits)	Clay, silt, fine-grained sand, organic rich soil, and peat	UC occurs primarily in the valleys occupied by Chimacum Creek and its tributaries (primarily East Fork, and typically is 20–50 feet thick (fig. 4A).
2	UA–Upper Aquifer unit (recessional outwash and till)	Sand, gravel, silt, and clay	UA occurs primarily to the west and south of the mouth of Chimacum Creek and typically is 30–60 feet thick, although quite variable (fig. 4B).
3	MC–Middle Confining unit (recessional outwash, till, and advance outwash deposits)	Unsorted and compacted clay, sand, and gravel; silt and clay	MC occurs primarily in the high elevations where the flutes from glacier passage are evident, and is absent in the valleys where it was likely scoured during glacial recession. It typically is 100–250 feet thick (fig. 4C).
4	LA–Lower Aquifer unit (till and advance outwash deposits)	Sand, gravel, silt, and clay	LA is roughly coincident with MC (also likely scoured during glacial recession), and typically is 50–200 feet thick (fig. 4D).
5	LC–Lower Confining unit (undifferentiated glacial and inter-glacial deposits)	Unsorted and compacted clay, sand, and gravel; silt and clay; lenses of sand and gravel	LC occurs primarily in the central peninsula, filling what can be thought of as a bedrock bowl deepening northward, and comprising a wide assortment of inter- and prior-glacial deposits. It typically is 100–300 feet thick, with large areas in the central peninsula 300–600 feet thick, and more than 1,000 feet thick at the northern extent of the model (fig. 4E).
6	OE–Bedrock unit (sedimentary and igneous rocks)	Sandstone, siltstone, shale, volcanic, and volcaniclastic rocks	OE occurs throughout the area and modeled thickness was between 268 and 2,407 feet (fig. 4F).

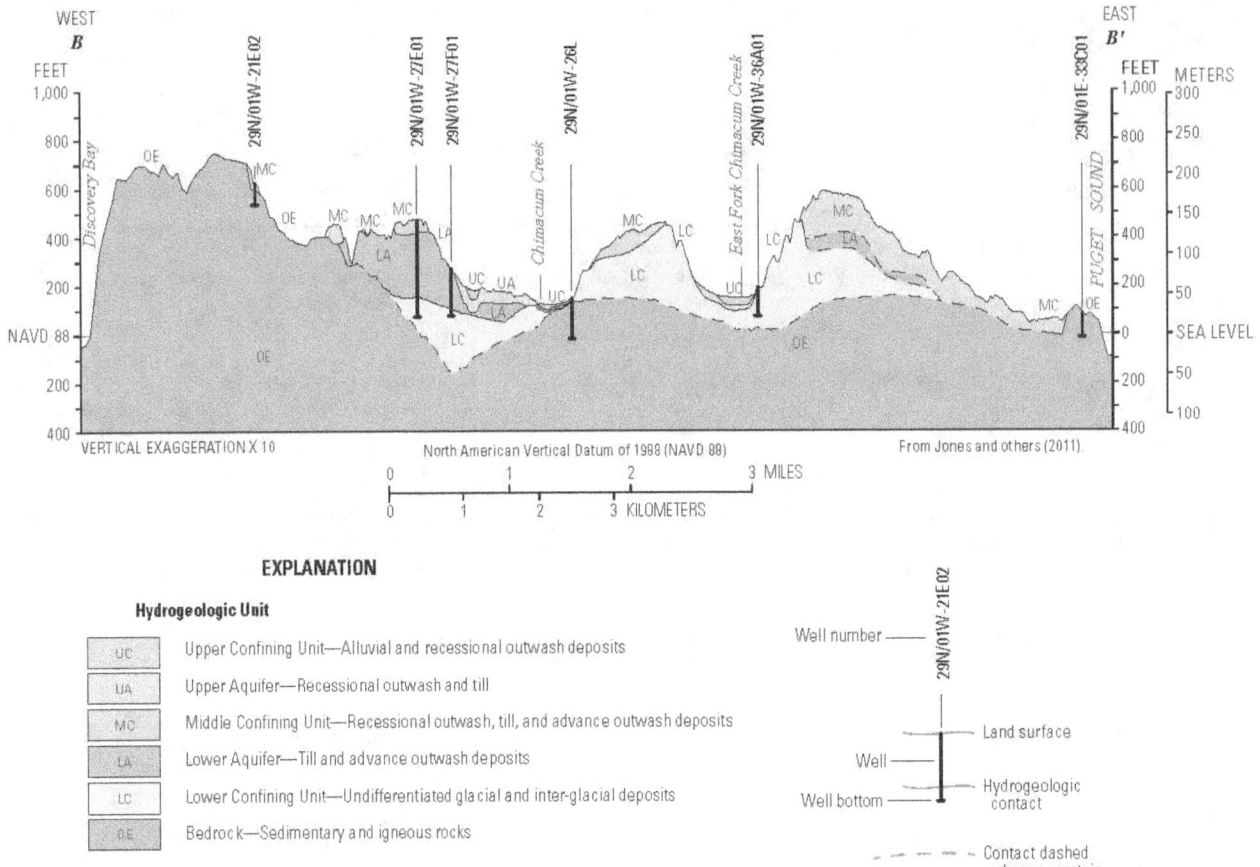

Figure 2. Example cross-section showing hydrogeologic units, Chimacum Creek Basin and vicinity, Jefferson County, Washington. (From Jones and others, 2011.)

Hydraulic Conductivity

Hydraulic conductivity is a measure of the ability of a material to transmit a given fluid, and in unconsolidated sediment is determined by the size, shape, distribution, and packing of individual particles. Because these characteristics vary greatly within each hydrogeologic unit, hydraulic conductivity values also vary greatly. Simonds and others (2004) estimated hydraulic conductivities for geologic units using drawdown data from drillers' logs. Median values were:

1. Vashon recessional outwash, 10 ft/d;

2. Vashon advance outwash, 130 ft/d;

3. Older glacial deposits, 22 ft/d, and

4. Bedrock, 0.53 ft/d.

Jones and others (2011) estimated hydraulic conductivities for hydrogeologic units using specific capacity data from drillers' logs for some of the monitoring wells (fig. 1):

1. Upper Aquifer (UA; roughly analogous to Vashon recessional outwash), 10 ft/d;

2. Lower Aquifer (LA; roughly analogous to Vashon advance outwash), 160 ft/d; and

3. Lower Confining (LC; analogous to Simonds and others' "older glacial") unit, 18.4–430 ft/d.

The estimates from Simonds and others (2004) and Jones and others (2011) are comparable to each other, and similar in magnitude to values described by Freeze and Cherry (1979) for similar materials. The large difference in values for LC from the monitoring wells is indicative of the great variety of lithologies in that hydrogeologic unit, which are not represented in model layer 5 due to lack of data available to make such distinctions.

Domestic water use by humans recharges groundwater by applying water to the land surface, where it evaporates, is taken up by plants, or percolates down to the water table.

Unused water that is pumped from any hydrogeologic unit before returning to the groundwater system is referred to as "return flow". Return flow may be defined as "The part of the water withdrawn for an agricultural, industrial or domestic purpose that returns to the groundwater or surface water in the same catchment as where it was abstracted. This water can potentially be withdrawn and used again." (Hoekstra and others, 2011). For indoor domestic use, the return flow (via septic systems) was estimated at 90 percent of water delivered for use, and for outdoor domestic use, 10 percent.

Surface-water "diversions"—withdrawals from Chimacum Creek or East Fork Chimacum Creek used exclusively for irrigation—also generate return flows that recharge the groundwater system. Surface-water return flows were assumed to be 10 percent of estimated surface-water irrigation use (including irrigation water from both groundwater and surface water sources). Water-use rates and return-flow estimates (table 2) used in this study were from Jones and others (2011). Recharge from all sources is presented in table 3.

Table 2. Water use in the Jefferson County Public Utility District #1 (public-supply use) and outside the public-supply area (self-supplied use) and recharge of groundwater by return flow from each class of user, during each year of the recorded period 1994–2009, with projections through 2030, Chimacum Creek Basin and vicinity, Jefferson County, Washington.

[Water year: The 12-month period October 1, for any given year through September 30, of the following year. The water year is designated by the calendar year in which it ends and which includes 9 of the 12 months. Thus, the year ending September 30, 1999, is referred to as water year 1999]

Water year	Source of usage or return flows, in acre-feet per year								
	Public utility district		Kala point	Self-supplied	Irrigation	Precipitation recharge	Public water return (Probable Use)	Total agriculture return	Self-supplied return
	Probable Use	Full Beneficial Use							
1995	622	1,406	172	179	327	18,261	567	76	121
1996	622	1,406	175	182	327	18,391	569	76	123
1997	501	1,406	176	183	327	19,476	487	76	124
1998	601	1,406	179	186	327	20,323	550	76	126
1999	597	1,406	181	189	327	23,108	566	76	128
2000	570	1,406	184	191	327	13,332	532	76	129
2001	599	1,406	189	197	327	13,055	559	76	133
2002	637	1,406	193	200	327	12,084	583	76	135
2003	695	1,406	195	203	327	13,297	623	76	137
2004	695	1,406	198	205	327	13,015	630	76	139
2005	665	1,406	201	209	327	13,727	608	76	141
2006	691	1,406	206	214	327	19,283	627	76	145
2007	688	1,406	208	216	327	13,956	631	76	146
2008	688	1,406	210	218	327	10,717	634	76	147
2009	751	1,406	211	219	327	13,132	669	76	149
2010	748	1,406	213	220	327	15,456	667	76	149
2011	759	1,406	215	223	327	15,456	676	76	151
2012	771	1,406	218	226	327	15,456	686	76	153
2013	783	1,406	221	229	327	15,456	697	76	155
2014	796	1,406	224	232	327	15,456	707	76	157
2015	809	1,406	226	235	327	15,456	718	76	159
2016	821	1,406	229	238	327	15,456	729	76	161
2017	833	1,406	232	241	327	15,456	739	76	163
2018	844	1,406	235	243	327	15,456	748	76	165
2019	855	1,406	237	246	327	15,456	758	76	167
2020	866	1,406	240	249	327	15,456	767	76	168
2021	878	1,406	242	251	327	15,456	777	76	170
2022	889	1,406	245	254	327	15,456	787	76	172
2023	901	1,406	248	257	327	15,456	797	76	174
2024	913	1,406	250	260	327	15,456	807	76	176
2025	925	1,406	253	263	327	15,456	817	76	178
2026	937	1,406	256	266	327	15,456	828	76	180
2027	950	1,406	259	269	327	15,456	838	76	182
2028	962	1,406	262	271	327	15,456	849	76	184
2029	975	1,406	265	274	327	15,456	860	76	186
2030	988	1,406	267	277	327	15,456	871	76	188

Table 3. Total estimable groundwater inflows (recharge) and outflows (discharge), and residual, Chimacum Creek Basin and vicinity, Jefferson County, Washington.

[Modified from Jones and others (2011)]

Recharge (acre-feet per year)	
From precipitation	15,600
From public supply return flows	592
From self-supplied return flows	136
From irrigated agricultural return flows	77
Total inflow to groundwater system	16,405
Discharge (acre-feet per year)	
Discharge to baseflow	[1]-8,174
Discharge to public supply	-833
Discharge to self-supplied domestic use	-200
Discharge to agricultural use	[2]-329
Total estimable outflow to groundwater only[2]	[2]-9,536
Residual (unobservable subaqueous outflow)	[2]6,869

[1]Modified from Jones (2011), which reported 4,600 acre-feet discharge to base flow based on three low-flow measurements; 8,174 acre-feet discharge to base flow is based on 84 measurements and flow separation.

[2]Modified from Jones and others (2011).

Groundwater-Flow Directions

Simonds and others (2004) used 110 field-verified wells and the measured land-surface altitudes to approximate groundwater table altitude and groundwater flow directions. The distribution of available wells for that study was limited by the low population density and large areas with limited accessibility or development in the study area. For areas with no data, Simonds and others (2004) relied largely on analysis of the topography, and assumed that radial flow moved away from topographic highs and northward down creek valleys. The western and eastern extents of the groundwater system are bounded by bedrock except for the area near the mouth of Chimacum Creek. East of East Fork Chimacum Creek, a topographic high suggests there should be a groundwater divide where groundwater either flows west to East Fork Chimacum Creek or east to discharge to springs, seeps or Oak Bay. A similar situation exists for the topographic high in the central part of the basin. The walls of the western and eastern valleys containing Chimacum Creek and East Fork Chimacum Creek have numerous springs discharging from areas where LA and coarser facies of LC units are exposed. Discharge from units is likely where UA or LA units are exposed to Discovery Bay or to Port Townsend Bay.

Little is known about vertical head gradients, but conceptually the confining units UC and LC are expected to contribute to a vertical gradient where they are bounded by aquifer units. Geohydrologic unit LC, however, is less consistently confining conceptually, and Simonds and others (2004, p. 23) note that

> "The lack of substantial confining layers in Qva [here, roughly LA] and the upper parts of Qgo [here, roughly LC] suggests that vertical ground-water [sic] movement between hydrogeologic units [sic, "geologic units" in the context of this report] is relatively uninhibited [sic, unimpeded]."

Groundwater Discharge

Groundwater discharges in the study area include well withdrawals and discharge to streams, which can be measured and estimated, and to springs and seeps located along the coastal bluffs, and submarine discharge to saltwater (Discovery Bay to the west, and Port Townsend and Oak Bays to the east). Coastal springs and seeps and submarine discharge are difficult to estimate, and generally are treated as a residual in any water budget. Groundwater withdrawals from wells during 1994–2009 averaged 833 acre-ft/yr for Group A Public Water Systems (PUD #1, B. Graham, unpub. data, 2009; Kala Point, Golder Associates, 2010), 200 acre-ft/yr for domestic wells (Jones and others, 2011), and 329 acre-ft/yr for agricultural wells (estimated from agricultural water use rights; Jones and others, 2011). These quantities represent gross withdrawals (self-supplied domestic and public water supply); they do not reflect the quantity of water returned to the groundwater system through septic systems, from outdoor domestic use, or from agricultural irrigation.

In addition to pumpage, groundwater discharges to Chimacum Creek, to its tributaries, and to springs. Spring discharges were not measured directly. Base flow near the mouth of Chimacum Creek was used as a surrogate for net base flow, and the study-estimated average base flow to be 11.29 ft³/s, or 8,174 acre-ft/yr, based on 84 streamflow measurements made by the WDOE and base flow separation by the hydrograph separation program HYSEP (Sloto and Crouse, 1996). Jones and others (2011) reported base flow as 6.36 ft³/s, or 4,600 acre-ft/yr based on three low flow measurements; table 3 is modified to reflect the HYSEP estimate. Total estimable outflows (discharges) to wells and streams is thus estimated to be 9,536 acre-ft/yr; the residual of 6,869 acre-ft/yr, is assumed to be unobservable discharge to seeps, springs, and submarine discharge (table 3).

Groundwater and Surface-Water Interactions

Gains and losses to and from stream channels display trends that are typically associated with local topography:

1. Gains occur in the channels incised into or through the till plain in the headwaters.

2. Losses occur along the long flat valley floors upstream of the confluence of the main stem and East Fork Chimacum Creek.

3. Gains occur where the streamflow is conveyed in a steeper and incised reach that carries it to Port Townsend Bay.

Simonds and others (2004) and Jones and others (2011) reported synoptic measurements that followed those general trends. Average annual base flow near the mouth of Chimacum Creek was 11.29 ft³/s, or about 8,174 acre-ft/yr, based on hydrograph separation using HYSEP (Sloto and Crouse, 1996).

Groundwater-Level Fluctuations

Monthly water-level measurements were made by PUD #1 during January 2008–May 2009 in eight monitoring wells (fig. 1) at the same locations that Simonds and others (2004) measured during 2002–03. Seasonal changes in groundwater levels that follow a pattern typical of shallow wells in western Washington were observed in most wells in both sets of data. Water levels typically rose in the autumn and winter and fell in the summer, although the timing of the seasonal signal varied from well to well. Most wells showed water level fluctuations on the order of 2–3 ft. The largest annual variations were at well 29N/01W-15R01 (about 13 ft for 2002–03 and 7 ft for 2008–09) and well 29N/01W-23F01 (about 8 ft for 2002–03 and 6 ft for 2008–09) (Jones and others, 2011). In both cases, the large variations are non-seasonal, large, short-lived water level declines in mid- to late summer, whereas the variations outside those periods are in the more typical 2 to 3 ft range, suggesting the possibility that the declines were related to anthropogenic withdrawals for agriculture (both wells are on the periphery of agricultural lands bordering the main stem of Chimacum Creek). Well 15R01 also is between two tributaries and near the periphery of the LA, possibly enhancing the effect of any anthropogenic withdrawals.

Groundwater Budget

Recharge from precipitation is the predominant inflow to the groundwater system for the Chimacum Creek Basin, averaging about 15,600 acre-ft/yr during 1994–2009. Outflows from the groundwater system include discharge to streams, withdrawals for domestic use and agriculture, and submarine groundwater discharge to saltwater bodies (table 3).

Most of the readily measurable discharge from the groundwater system is discharge to streams as base flow. Stream base flow during 2003–09 averaged 11.29 ft³/s, or 8,174 acre-ft/yr. Groundwater is the sole source of domestic use. Jefferson County PUD #1 provides most of the water for domestic use, averaging about 833 acre-ft/yr. Of this, 90 percent of indoor usage is considered to recharge the groundwater system by way of septic systems, and 10 percent of outdoor use is considered to recharge the groundwater systems, resulting in about 70 percent, or about 592 acre-ft of discharge per year returning to the groundwater system. The remaining domestic water use is self-supplied, which amounts to just less than 200 acre-ft, of which 136 acre-ft (about 70 percent of withdrawals) is returned to the groundwater system. Agricultural groundwater use was estimated at about 329 acre-ft (Golder Associates, 2010), and 10 percent of that is considered to recharge to the groundwater system.

Balancing mean annual groundwater recharge and discharge, of the estimated 16,405 acre-ft of recharge to the system (natural recharge plus return flows), base flow (8,174 acre-ft), public water supply (833 acre-ft), self-supplied water (200 acre-ft), and agricultural water use (329 acre-ft), account for 9,536 acre-ft. The remainder, 6,869 acre-ft, is assumed to be immeasurable discharge to springs and seeps, and submarine groundwater discharge to surrounding saltwater bodies.

Numerical Simulation of Groundwater-Flow System

Groundwater flow in the Chimacum Creek Basin and vicinity was simulated using the U.S. Geological Survey modular three-dimensional finite-difference groundwater-flow model, MODFLOW-2005 (Harbaugh, 2005). The model described in this report was developed to simulate steady-state conditions. Steady-state groundwater flow represents a groundwater system that is in a state of equilibrium: inflows into and outflows from the system are constant and equal, resulting in no changes in groundwater storage. No long-term ambient groundwater monitoring network is in the model area, and data from a short-term monthly monitoring network established by Simonds and others (2004) and reestablished for this study (Jones and others, 2011) are insufficient to evaluate water-level trends relating to long-term changes in groundwater storage, or for testing the assumption of steady-state conditions. Steady-state conditions were approximated by selecting a period that spanned several years, and thereby included complete seasonal cycles. Although there was some change during this time, the flow discrepancies due to changes in storage were minimized by the length of time involved. The calibration to steady state conditions incorporated average annual values for recharge from October 1994 through September 2009, discharge, and other groundwater-flow system processes.

Model Grid and Layering

The MODFLOW program simulates the conceptual model of the groundwater-flow system using data sets that describe the hydrogeologic units, and estimates of recharge and discharge, to calculate hydraulic heads at discrete points, and flow within the model domain. The program requires that the groundwater-flow system be subdivided, vertically and horizontally, into model cells. The hydraulic properties of the material in each cell are assumed to be homogeneous. The Chimacum Creek study area was subdivided by a horizontal grid of 245 columns and 313 rows; cells are a uniform 200 ft per side (fig. 3). The cell size and uniform grid spacing were selected to reflect the regional scale of this study. The extents and thickness of the active cells in each layer are outlined in figures 4A–F.

The hydrogeologic units delineated by Jones and others (2011) in table 1 were used to represent the three-dimensional hydrogeologic framework in the model. Most units do not extend across the entire model area, and unit thicknesses vary considerably over short distances. Vertically, the study area was subdivided into six layers having varying thicknesses (fig. 2). Each hydrogeologic unit is represented by a single model layer (table 1) that closely corresponds to the stratigraphic position of the unit in the aquifer system based on the unit top and thickness. Five model layers (fig. 4A–E) were used to simulate the saturated unconsolidated sediments that overlie the bedrock, and one layer was used to simulate the upper bedrock (fig. 4F) down to an altitude of 1,500 ft below NAVD 88.

Although parts of model layers 1–3 were conceptually unconfined, all model layers were simulated as confined so that the transmissivity of each cell remained constant throughout the duration of the simulation. This simplification greatly improved the numerical stability of the model. Provided the percentage change in head in response to forecast changes in recharge or discharge is less than around 10 percent, this treatment of the conceptually unconfined parts of a groundwater system is generally considered accurate (Faunt and others, 2011). Furthermore, the difference between the percentage change in head using the model layer thicknesses as the saturated thicknesses (confined flow case) versus using smaller saturated thicknesses (unconfined flow case) was expected to be negligible. To account for areas where the hydrogeologic units constituting a model layer were absent, the layer was assigned a 5-ft thickness, and the hydraulic properties were specified to be the hydraulic conductivities of the underlying unit present at that location; this also was done for any intermediate absent layer. This procedure resulted in the simulated flow passing through the absent layer as if it were part of an adjacent model layer.

Time Discretization

The calibration to steady state conditions used average conditions for the period of October 1994–September 2009. This calibration estimated the hydraulic conductivities of all layers, conductances for most boundary conditions, and preliminary stream conductances. Streambed conductance is the volumetric discharge of water passing through the streambed per foot of head gradient between the stream and the aquifer and is defined as the hydraulic conductivity of the streambed times the area of the streambed, divided by the streambed thickness. Stream conductances can rarely be measured or estimated over long reaches of streams; in practice, they are determined during calibration.

Boundary Conditions

Boundary conditions in a groundwater-flow model are used to specify where water enters and exits the active model domain. For the general conceptual model for the Chimacum Creek Basin model, water enters the system as recharge from precipitation and return flows (septic-system and irrigation returns) to the water table and exits the system as groundwater pumpage, groundwater discharge to streams and springs, seeps along the marine bluffs, and submarine discharge to surrounding saltwater bodies. The boundaries of the model coincide with natural topographic, geologic, and hydrologic boundaries except the northern edge, which was located as far north as possible without approaching areas likely affected by return flows and groundwater withdrawals from the Port Townsend urban area. Three types of boundaries were used in the model: specified flux (recharge and pumping wells), head-dependent flux (constant head, general head, and drains), and no flow (outer model boundary) (fig. 3). The bottom boundary of the model is a no-flow boundary (bottom of layer 6 at an altitude of -1,500 ft). The areal boundaries along the southern edge of the model correspond with the drainage basin boundaries of Chimacum Creek (fig. 3). These natural features act as no-flow boundaries as they are considered coincident with groundwater divides.

Model layer 1 (figs. 3 and 4A) of the model includes specified-flux and head-dependent-flux boundary cells. The specified-flux boundary is areally applied groundwater recharge, and the head-dependent boundaries represent streams, springs, or groundwater seeps. Recharge was simulated with the recharge (RCH) package (McDonald and Harbaugh, 1988). Streams, springs, and groundwater seeps were simulated with either the drain (DRN) package (McDonald and Harbaugh, 1988) or the stream (STR) package (Prudic, 1989).

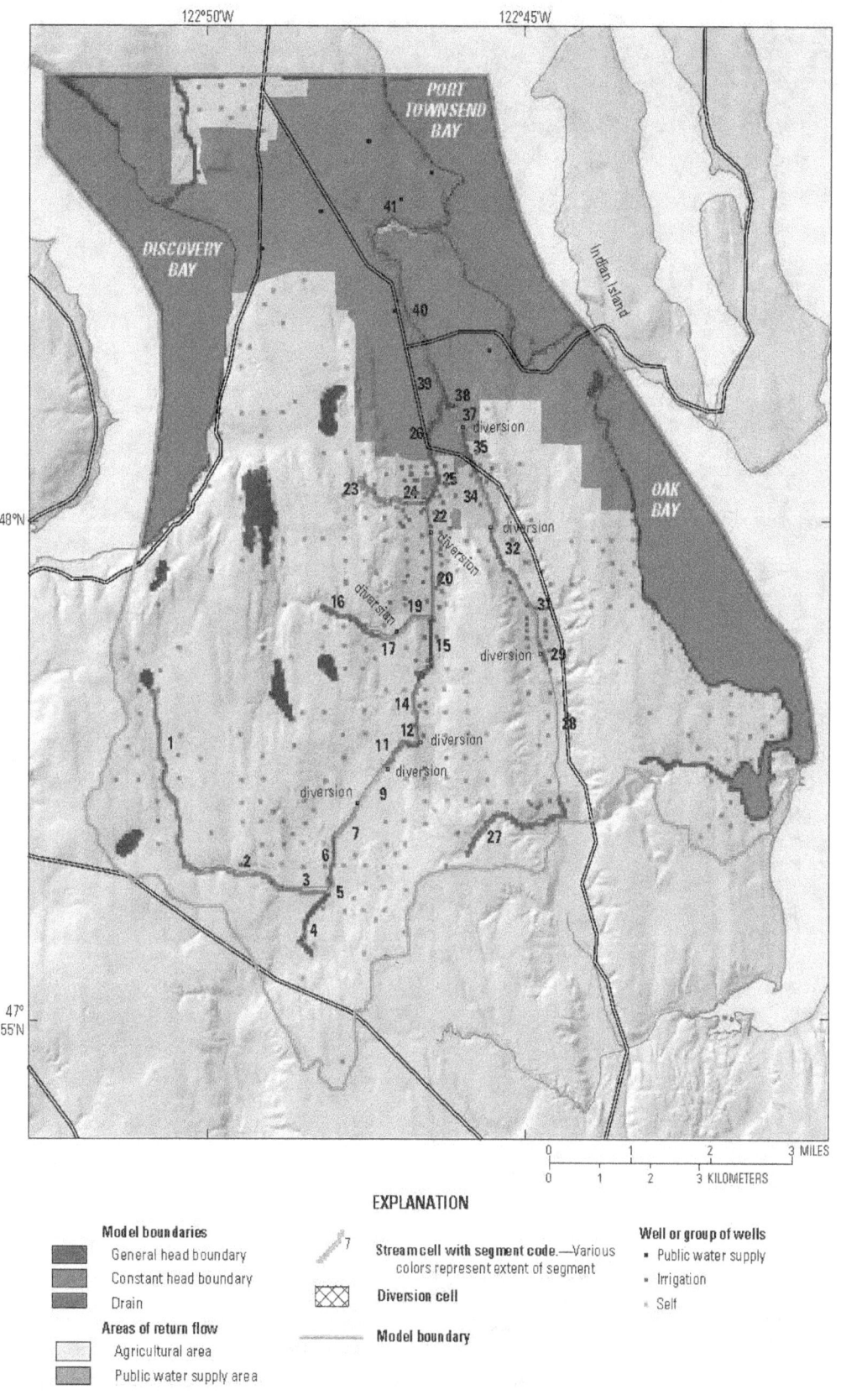

Figure 3. Location of model boundary condition cells, Chimacum Creek Basin and vicinity, Jefferson County, Washington.

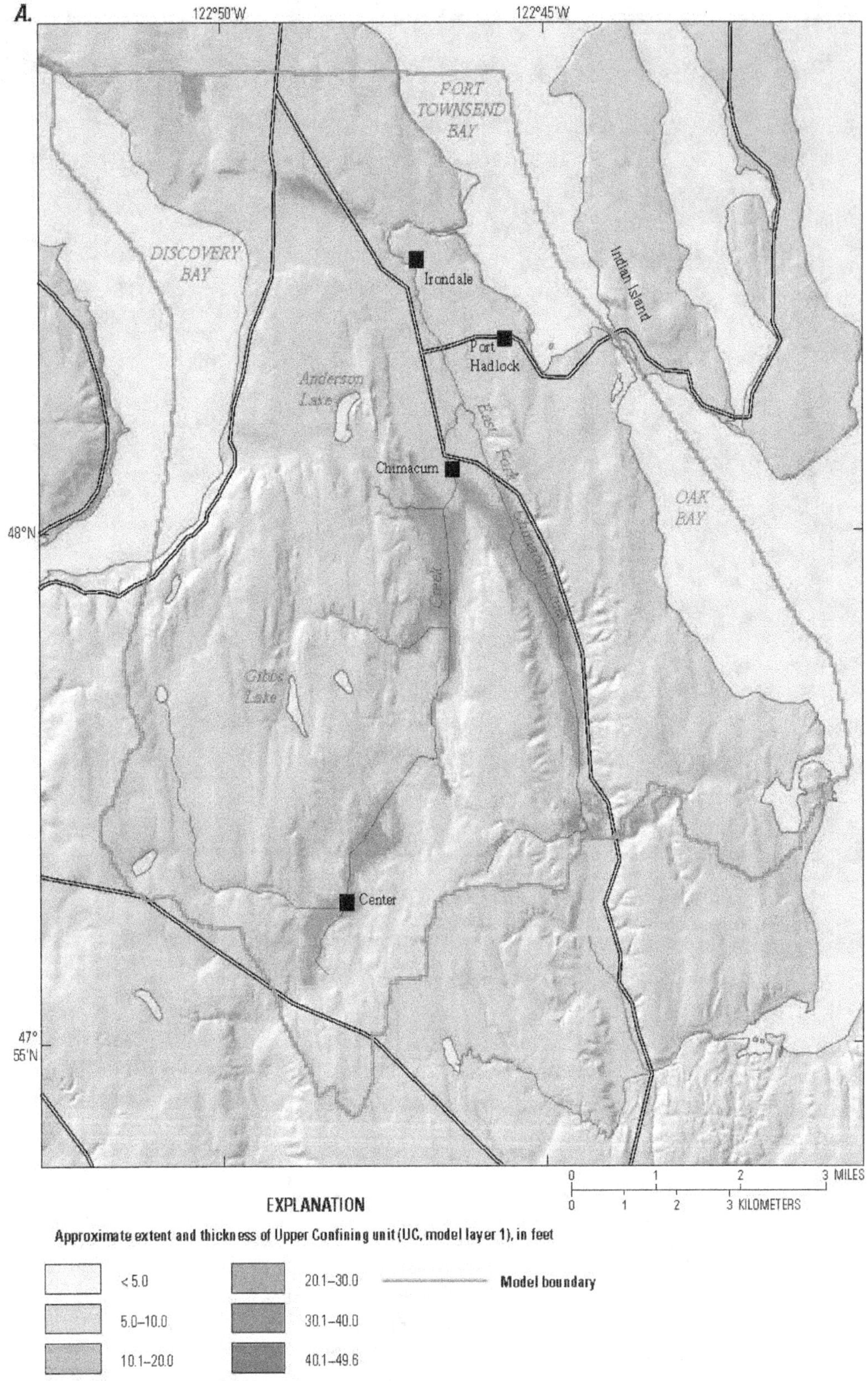

Figure 4.   Extent and thickness of model layers, Chimacum Creek Basin and vicinity, Jefferson County, Washington.

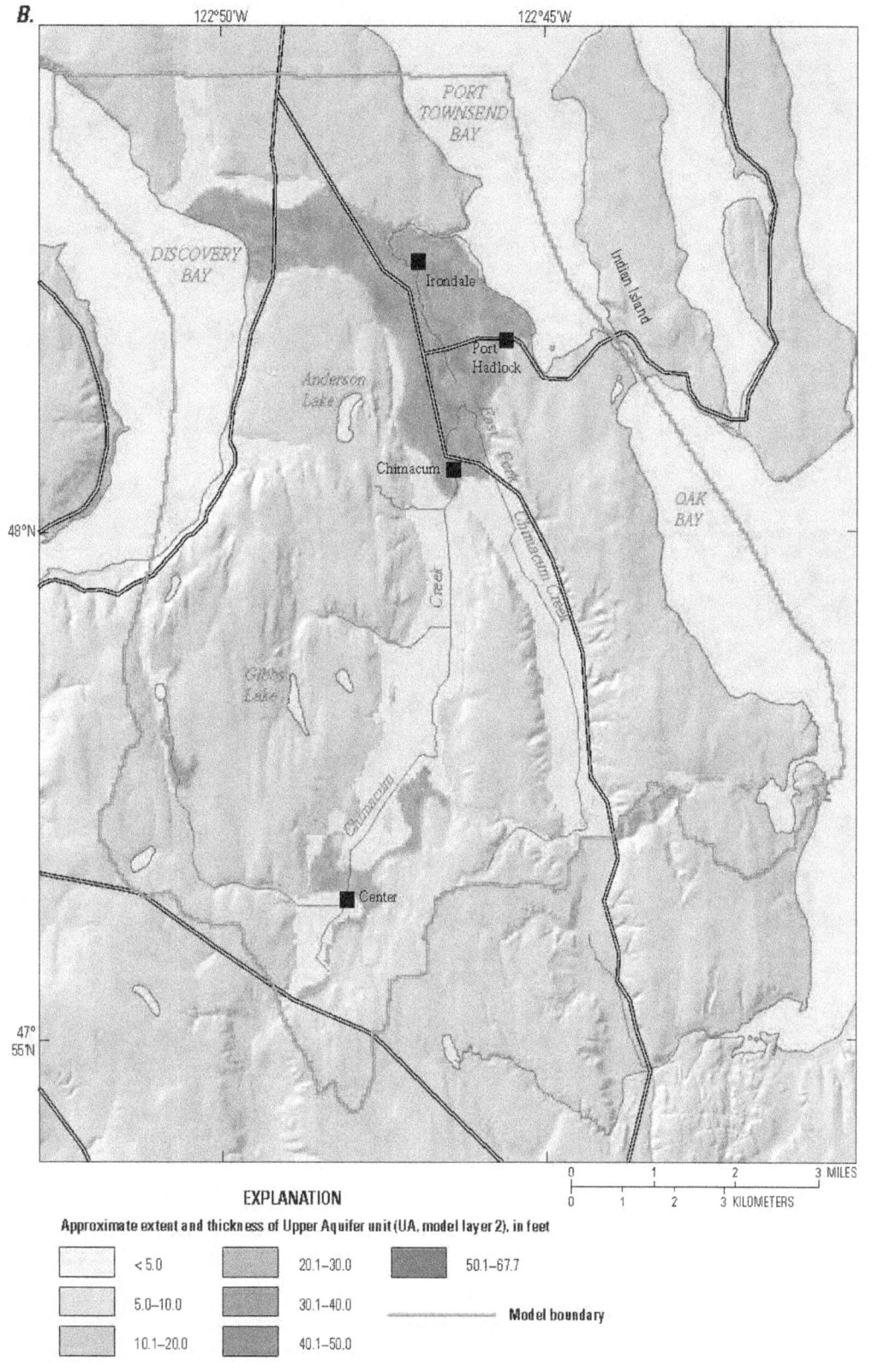

B.

EXPLANATION

Approximate extent and thickness of Upper Aquifer unit (UA, model layer 2), in feet

< 5.0	20.1–30.0	50.1–67.7
5.0–10.0	30.1–40.0	Model boundary
10.1–20.0	40.1–50.0	

Figure 4.—Continued

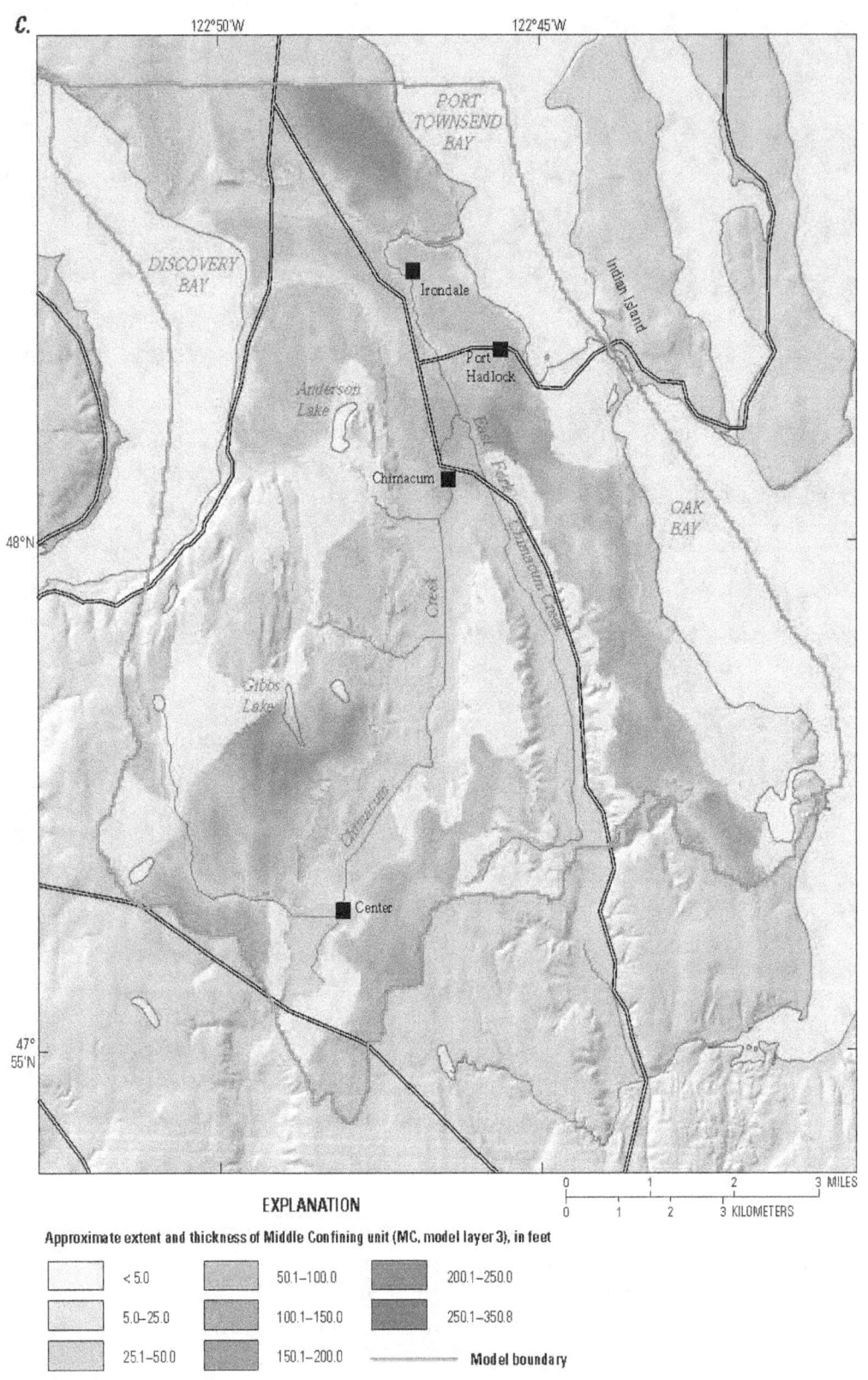

EXPLANATION

Approximate extent and thickness of Middle Confining unit (MC, model layer 3), in feet

< 5.0	50.1–100.0	200.1–250.0
5.0–25.0	100.1–150.0	250.1–350.8
25.1–50.0	150.1–200.0	Model boundary

Figure 4.—Continued

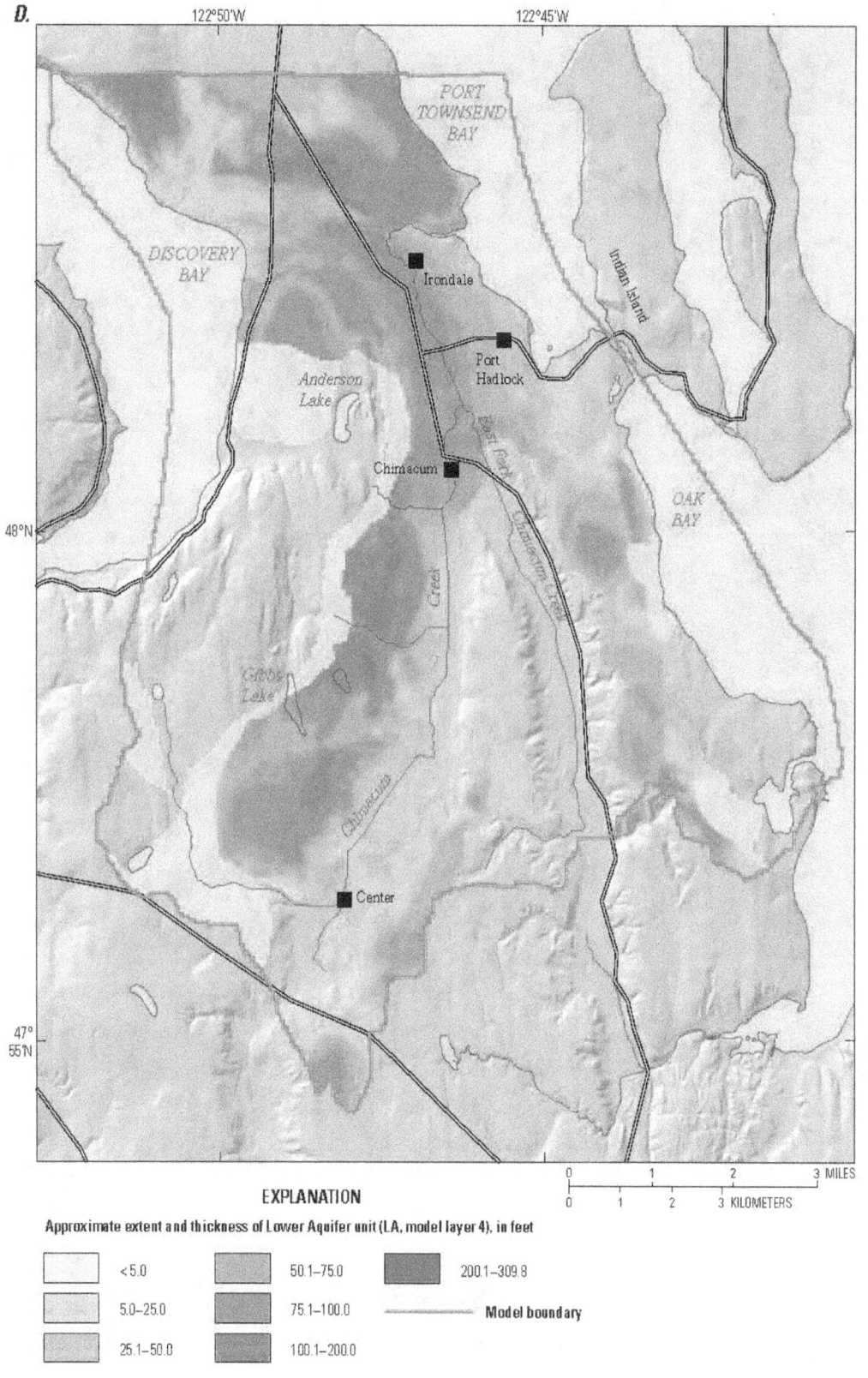

EXPLANATION

Approximate extent and thickness of Lower Aquifer unit (LA, model layer 4), in feet

< 5.0	50.1–75.0	200.1–309.8
5.0–25.0	75.1–100.0	Model boundary
25.1–50.0	100.1–200.0	

Figure 4.—Continued

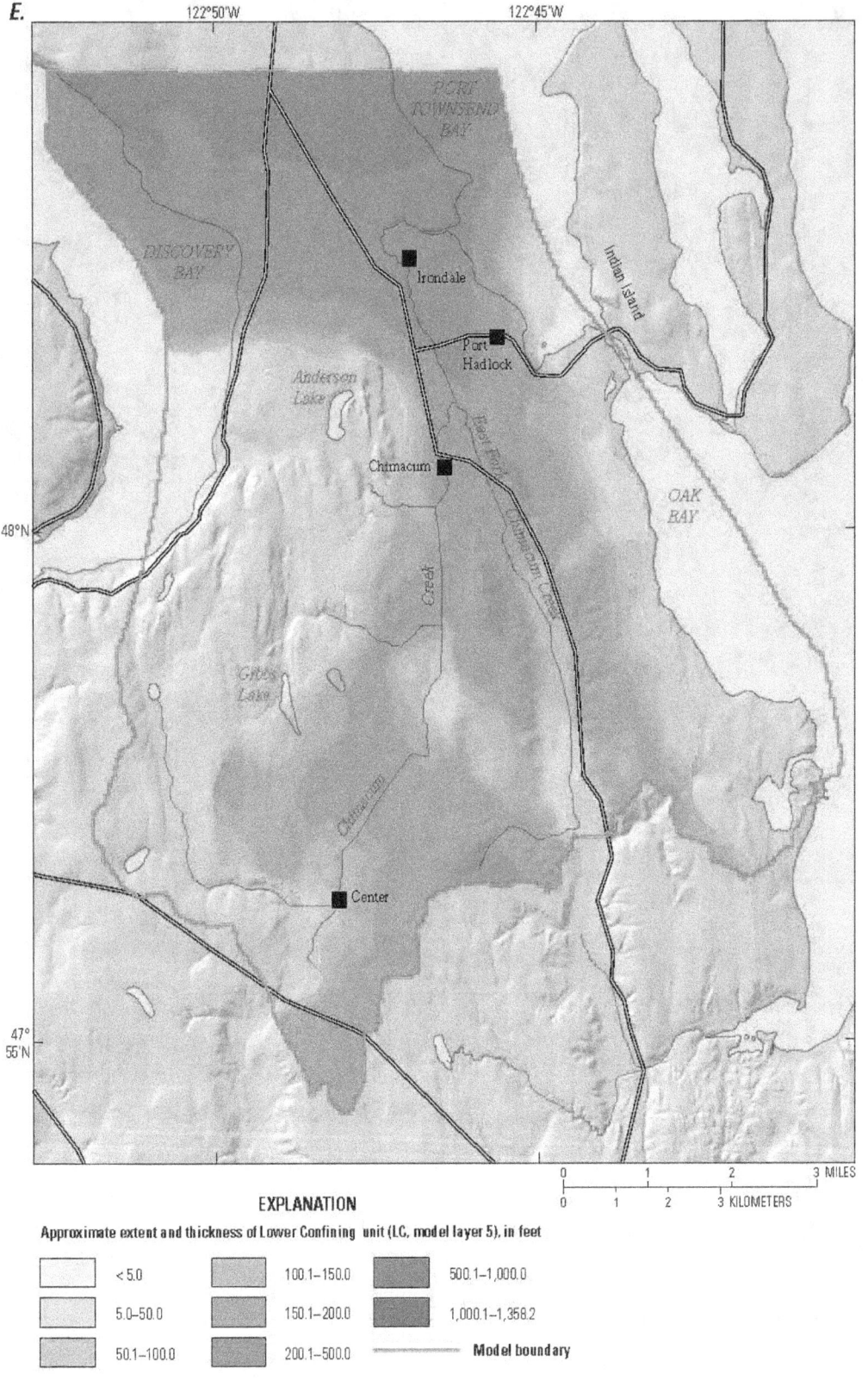

EXPLANATION

Approximate extent and thickness of Lower Confining unit (LC, model layer 5), in feet

< 5.0	
5.0–50.0	
50.1–100.0	
100.1–150.0	
150.1–200.0	
200.1–500.0	
500.1–1,000.0	
1,000.1–1,358.2	
Model boundary	

Figure 4.—Continued

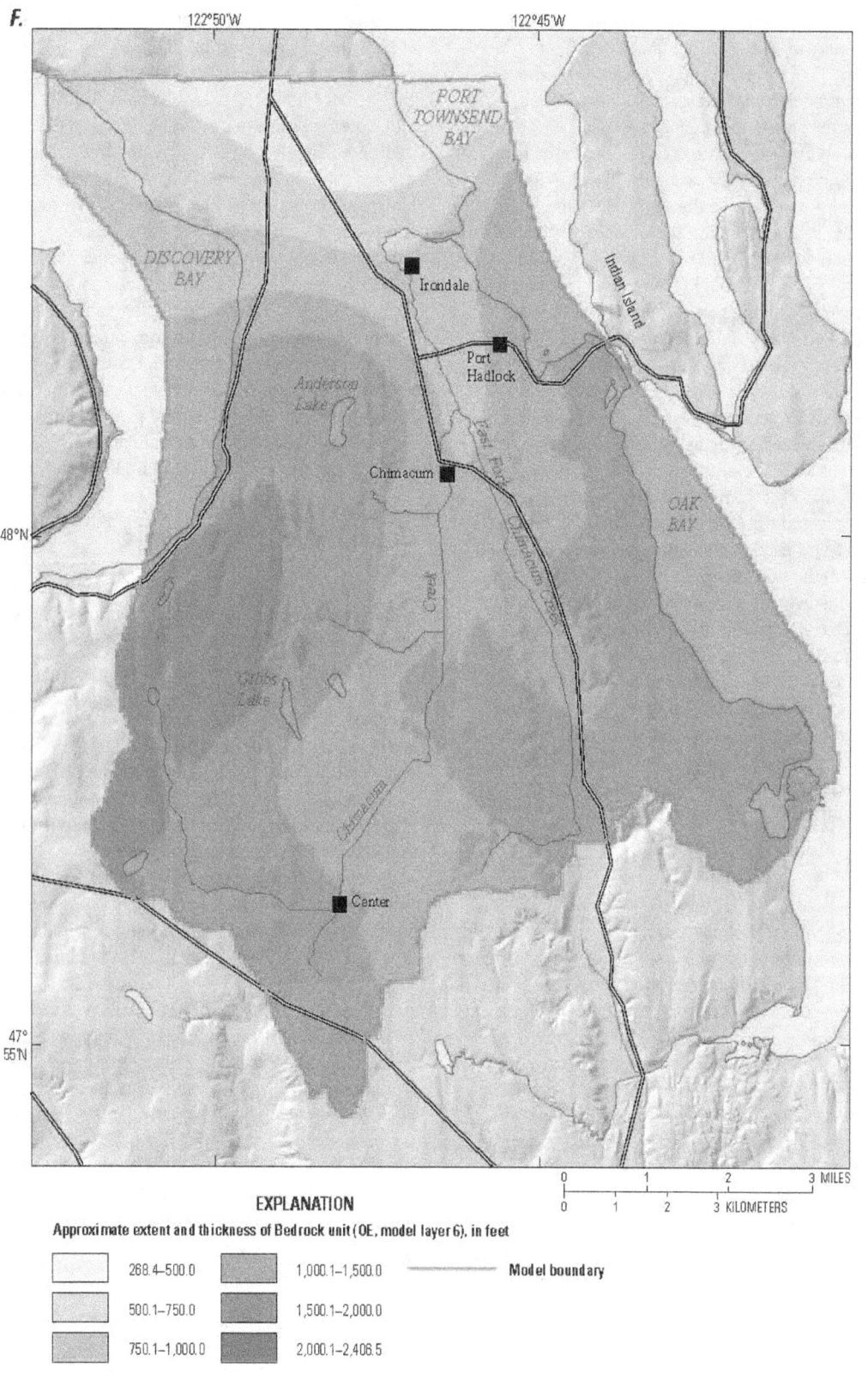

EXPLANATION

Approximate extent and thickness of Bedrock unit (OE, model layer 6), in feet

268.4–500.0		1,000.1–1,500.0	——— Model boundary
500.1–750.0		1,500.1–2,000.0	
750.1–1,000.0		2,000.1–2,406.5	

Figure 4.—Continued

Offshore submarine boundaries along the periphery of the model in layers 2–6 were simulated as general head boundaries (fig. 3). In layer 1 (unit UC), cells in contact with saltwater were modeled as constant-head boundaries (fig. 3). The head assigned to offshore boundaries was the altitude of the freshwater-equivalent head corresponding to the height of the saltwater column above the seafloor (Bear, 1979).

The northern boundary of the model was simulated as head-dependent-flux boundaries in model layers 2 and 4 (aquifer units UA, roughly corresponding to Vashon Recessional Outwash; LA, roughly corresponding to Vashon Advance Outwash) with the general-head-boundary (GHB) package of MODFLOW (McDonald and Harbaugh, 1988). This allows groundwater to enter and exit the northern boundary from the area to the north. Fluxes into the model in these two aquifer layers were estimated from the area of surface-water drainage flowing into the area from the north and used as targets in the calibration process. Model layers 1 (unit UC), 3 (unit MC), 5 (unit LC), and 6 (unit OE, bedrock) have no-flow boundaries. The GHB package of MODFLOW was used to simulate subsurface discharge to the lakes from the underlying aquifers. Representation of the lake in this way allows flow into and out of a cell in proportion to the difference between the head in the cell and the specified head of the lake. The specified lake stages were determined from USGS 1:24,000-scale topographic maps. Where thin unconsolidated material overlies bedrock, such that recharge is calculated as non-zero due to the presence of the unconsolidated material, drains were used to lower heads that built up to unreasonable values due to the low hydraulic conductivity of the underlying bedrock.

Recharge

Precipitation is the dominant source of water recharging the groundwater system in the study area, and the distribution of recharge from precipitation in the study area (fig. 5) was estimated by Jones and others (2011) using the precipitation-recharge regression equations of Bidlake and Payne (2001). Recharge to the area from septic-system return flows were estimated as previously described in Jones and others (2011). The return flows in the public water-supply distribution areas and agricultural areas (fig. 3) were applied to each corresponding model cell in layer 1 using the MODFLOW recharge package RCH. Return flows from domestic septic-system and outdoor irrigation in the public water supply area were distributed evenly over the service area and merged with the recharge from precipitation using a Geographic

Information System; for septic systems outside the service area—where water is from domestic wells—the return flows were simulated as injection wells in layer 1 using the WELL package of MODFLOW (McDonald and Harbaugh, 1998) and were located using the WDOE well database. Where hydrogeologic unit UC, represented by model layer 1, is absent, it was simulated as a thin (5 ft) layer and assumed the hydraulic properties of the uppermost hydrogeologic unit that is present.

The initial calibration of the model used monthly values averaged over October 1994 through September 2009 for precipitation recharge, pumping, and return flows. For simulations of possible future conditions, a "Current Conditions" version of the model was developed from that calibration using National Oceanic and Atmospheric Administration defined "normal precipitation," (National Oceanic and Atmospheric Administration, 2013) and current (2009) precipitation recharge, pumpage, and return flows.

Groundwater Withdrawals

The WELL package was used to simulate groundwater withdrawals from more than 700 pumping wells (fig. 3, shown at, and many consolidated to, township-range quarter-quarter centers). The WELL package simulates a specified-flux boundary in each model cell to which a well is assigned based on the withdrawal rate for each well or group of pumping wells in the cell. The reported depth of the open interval of each well was used to determine the model layer from which the withdrawal was made. Actual monthly pumping totals (fig. 6) were obtained from the largest public-supply systems, and domestic-pumping amounts were estimated using the WDOE well database (Washington State Department of Ecology, 2003) and the monthly per-capita rates from Golder Associates (2008). For calibration, the withdrawals were averaged for October 1, 1994, to September 30, 2009 (water years 1995–2009).

Pumping from agricultural wells was estimated by subtracting the total amount of irrigation water taken from Chimacum Creek, based on agricultural water rights permits, from the total amount of water estimated to be used for agricultural purposes in the basin (770 acre-ft/yr; Golder Associates, 2010). The difference, 329 acre-ft/yr, was assigned to wells in layer 4 (unit LA) that were assumed to exist on each irrigated agricultural parcel. For steady-state simulations, agricultural groundwater withdrawals are effectively distributed over the year evenly (the entire amount was applied as an annual stress).

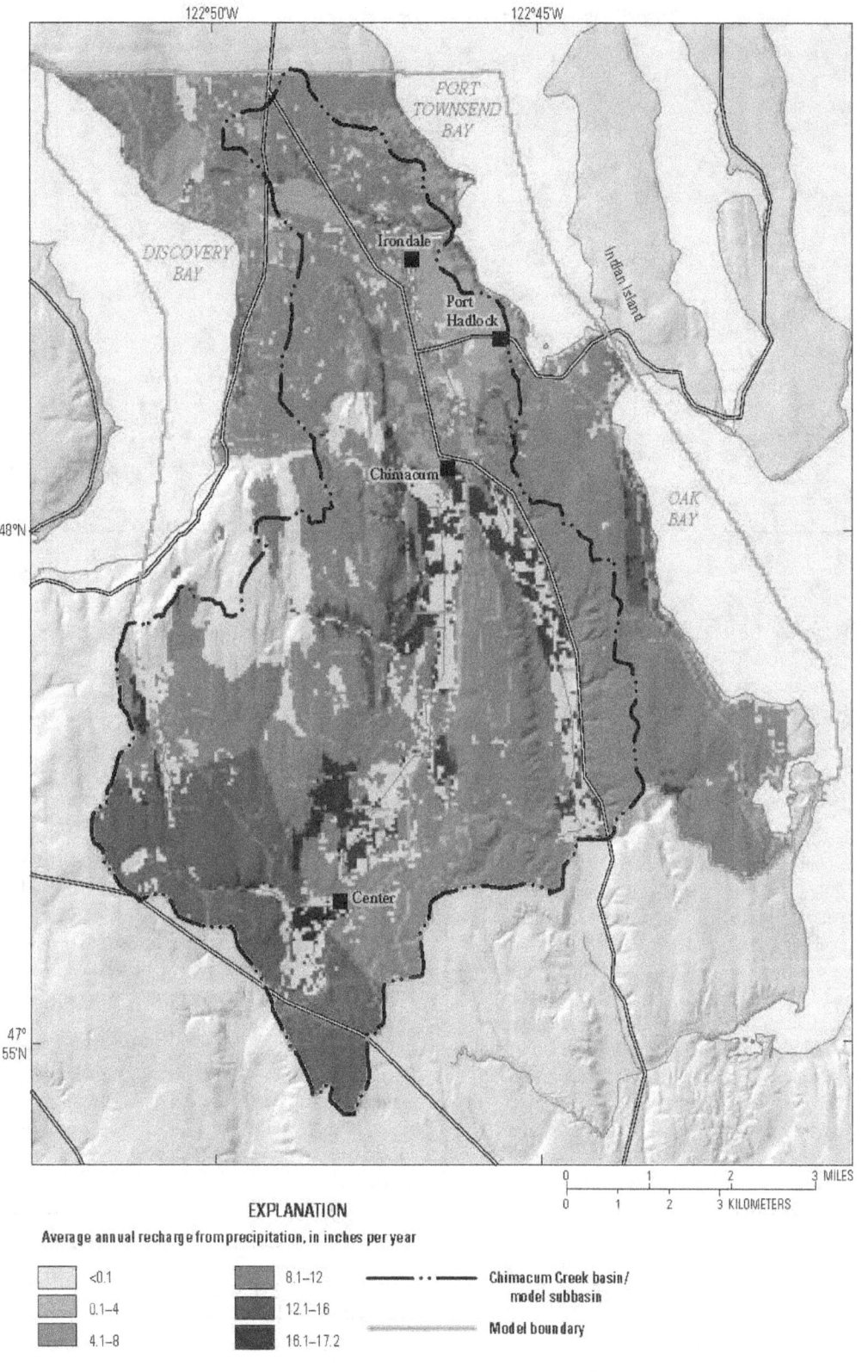

Average annual recharge from precipitation, in inches per year

<0.1	8.1–12	Chimacum Creek basin / model subbasin
0.1–4	12.1–16	Model boundary
4.1–8	16.1–17.2	

Figure 5. Average annual recharge from precipitation in the Chimacum Creek Basin and vicinity, Jefferson County, Washington, water years 1995–2009. (From Jones and others, 2011.)

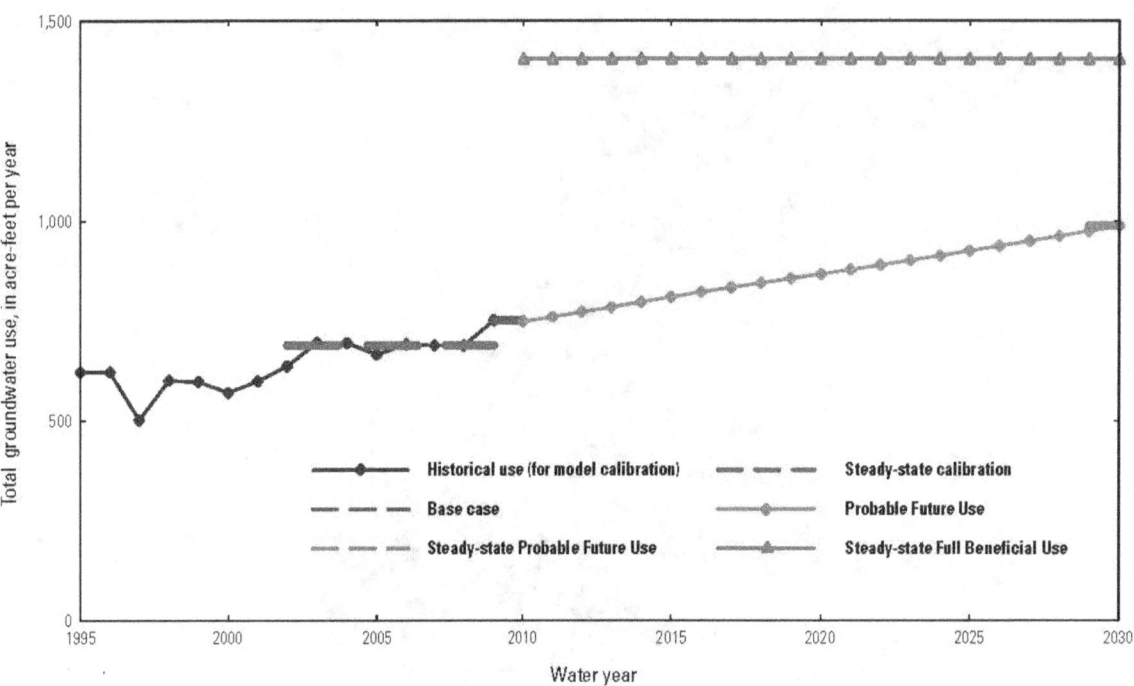

Figure 6.   Historical groundwater use, probable future groundwater use, steady-state full beneficial groundwater use and total groundwater use for Public Utility District #1, Chimacum Creek Basin and vicinity, Jefferson County, Washington, water years 1995–2030.

Streams

Chimacum Creek was simulated using the STR Package (Prudic, 1989) of MODFLOW, which allows water to flow from the groundwater system to the stream or conversely, depending on the relative stream and groundwater levels. If the specified stream stage is lower than the simulated groundwater level in the cell, water will discharge from the groundwater-flow system to the stream. The reverse happens if the assigned stage in the stream cell is higher than the simulated water level in the cell. The rate at which the recharge to or discharge from the groundwater-flow system occurs depends on the magnitude of the water-level difference and the streambed conductance. The conductance was assigned to each cell within an STR stream segment of the stream (fig. 3) through the calibration process. The STR package sums the groundwater discharge to the stream from upstream to downstream, which allows comparison to streamflow records at the WDOE streamgage that serves as a regulatory point (17B050, fig. 1). Discharge from the stream to the flow system is possible as long as water remains in the stream, which is known to be perennial.

In the model, stream stages were estimated using LiDAR-derived altitudes along the stream channel. Each assigned stage was taken from the center of the stream reach for each cell. Streams in the study area were simulated as 1-ft deep. Diversions from the streams (fig. 3) were calculated based on the locations of water rights permits obtained from the WDOE (Dave Nazy, Washington State Department of Ecology, written commun. 2010). The irrigated acreage for each permit was multiplied by an estimated water application rate of 1.39 (acre-ft/acre)/yr (Golder Associates, 2010) and was assumed to remain constant each year. Ten percent of this amount was applied to layer 1 as return flow in the areas corresponding to the water rights.

Tributaries to Chimacum Creek were simulated using the DRN Package of MODFLOW, which allows water to flow from the groundwater system to the streams if the simulated groundwater level in the model cell is greater than the specified altitude of the stream in the drain cell. The altitude of the drain cell was set equal to the altitude of the centroid of the stream segment present within a given model cell. DRN was selected to represent these small streams because the streams in the study area are groundwater-fed and could dry up in the summer if the water table declines below the altitude of the streambed due to forecast increased groundwater use. Drains also were used to represent seepage to springs and groundwater seeps that occur on the periphery of the area in bluffs along the coast. The altitude of these drain cells was set to the land-surface. Drains also were placed in areas where bedrock beneath thin unconsolidated sediments resulted in unreasonably large head values (the drain discharges are accounted for in the springs/seeps part of groundwater budgets). The altitude of these drain cells was set to the land-surface altitude.

Model Calibration

The model was calibrated assuming steady-state conditions to estimate hydraulic conductivities of the six hydrogeologic units to match observed heads and flows at target wells and streamflow sites (fig. 7). Average values of observed heads and flows from October 1994 through September 2009 were compared to simulated heads and flows. Parameters estimated included (1) horizontal and vertical hydraulic conductivity of aquifers and confining units; (2) stream conductances for 33 stream reaches (all reaches except those with diversions); (3) GHB conductances for the groundwater inflow along the northern boundary, for the lakes, and for submarine discharge areas; and (4) drain conductances for smaller streams and along coastal bluffs.

The model calibration was done using nonlinear regression with the parameter-estimation program (PEST), with regularized inversion (Doherty, 2003; 2005), pilot points to represent heterogeneity of aquifer and confining unit properties, and Singular Value Decomposition Assist (SVDA). This approach allowed a relatively large number of parameters (189 parameters) to be estimated using a set of pilot points distributed throughout the model domain (Doherty, 2003; 2005). Numerous studies have described the use of pilot points for groundwater model calibration (de Marsily and others, 1984; LaVenue and Pickens, 1992; Petkewich and Campbell, 2007). Hydraulic properties of each hydrogeologic unit within the model were estimated through spatial interpolation using kriging from the pilot points to the model grid cells. The result was a smooth variation of the hydraulic property values within each unit of the model domain. Twenty pilot point locations (fig. 8) used for hydraulic conductivity were evenly spread throughout the model domain and were distributed vertically so that each hydrogeologic unit contained pilot points. Pilot points were not specified where a hydrogeologic unit was absent, resulting in 150 active hydraulic conductivity pilot points (75 for horizontal hydraulic conductivity, Kx, and 75 for vertical hydraulic conductivity, Kz).

Observations were weighted differently to reflect the uncertainty of the measured values. Different weighting factors were used for the water-level and flow measurements to ensure equal contributions of each type of observation in the nonlinear regression.

The steady-state calibration used average water-level measurements in 57 wells (table 4) and average surface-water discharge measurements at 13 locations (table 5): the streamgage 17B050 (synoptic seepage base flow location CS14) and 12 other synoptic seepage base flow locations (CS1, and CS3–CS13), plus estimated flows in aquifer layers 2 and 4 through the northern boundary of the model. The reported depth of the well screen and the well log were used to determine the model layer that represented the hydrogeologic unit screened by the well; for the small number of wells screened in multiple units, the unit with the larger screened interval was assigned. Surface-water base flow measurements were collected at 13 sites (fig. 7) during 3 synoptic events (June and October 2002 and July 2007) that represent periods of low flow. Mean monthly base flow values were also computed from the streamflow data measured at the WDOE streamgage 17B050 near the mouth of Chimacum Creek for October 2002–September 2009 (water years 2003–09).

Sensitivity Analysis and Final Parameter Values

The sensitivity of the simulated model output to changes in a parameter value determines the uncertainty of the estimated parameter values; values are better estimated for parameters with a high sensitivity (a large effect on simulated head). In contrast, changing the value of parameters with low sensitivity has little effect on the model-calibration process, and values for these insensitive parameters are not well estimated. Values for 189 parameters (that is, pilot points for Kx and Kz, and conductances for general head, drain, and stream boundaries) were computed in the steady-state calibration. Sensitivities for these parameters were calculated using an "identifiability" measure that is included in the PEST procedure (Doherty and Hunt, 2009) based on "singular values" and associated vectors that are part of the SVDA procedure. These vectors relate each parameter to its influence on the objective function that is the sum of squared weighted errors at all the target wells and streamflow targets. The resulting identifiability values (the square root of the sum of the vector components for a given parameter) are shown in figure 9 where higher identifiabililty values indicate higher sensitivity. The model is most sensitive to horizontal conductivities (Kx) in layers 2, 4, and 6 with pilot points in the central part of these units having the highest identifiability values.

The final values for calibration parameters are listed in table 6 and the areal distribution of horizontal and vertical conductivities are shown in figures 8A–L. The properties of layer 2 (UA) are the dominant variables (hydraulic conductivity both vertical and horizontal), and the properties of layer 4 (LA) are similarly significant.

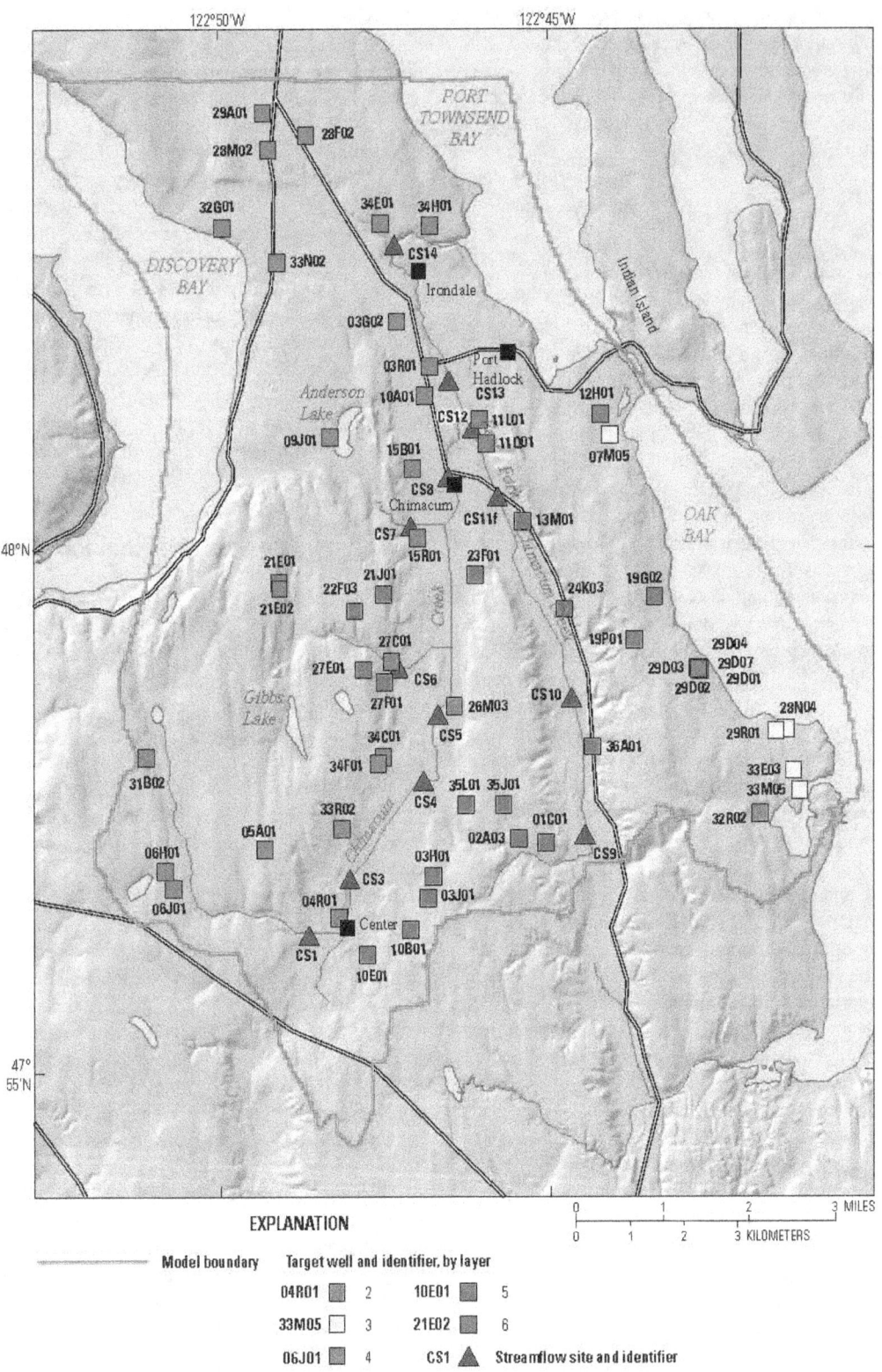

Figure 7.   Locations of wells and streamflow sites used for calibration, Chimacum Creek Basin and vicinity, Jefferson County, Washington.

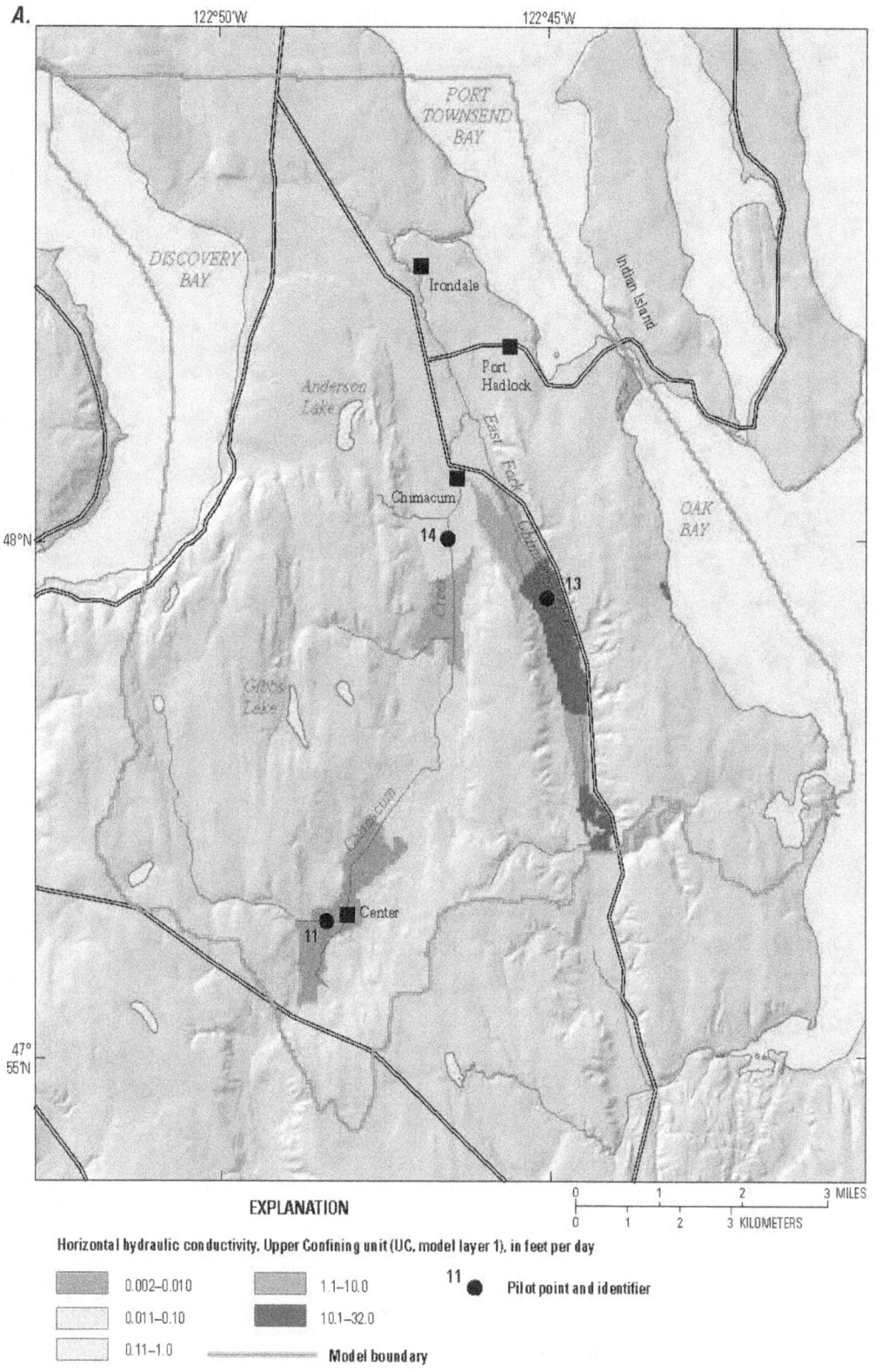

Figure 8. Aerial distribution of horizontal (*A-F*) and vertical conductivities (*G-L*) and locations of hydraulic conductivity calibration pilot points for all model layers, Chimacum Creek Basin and vicinity, Jefferson County, Washington.

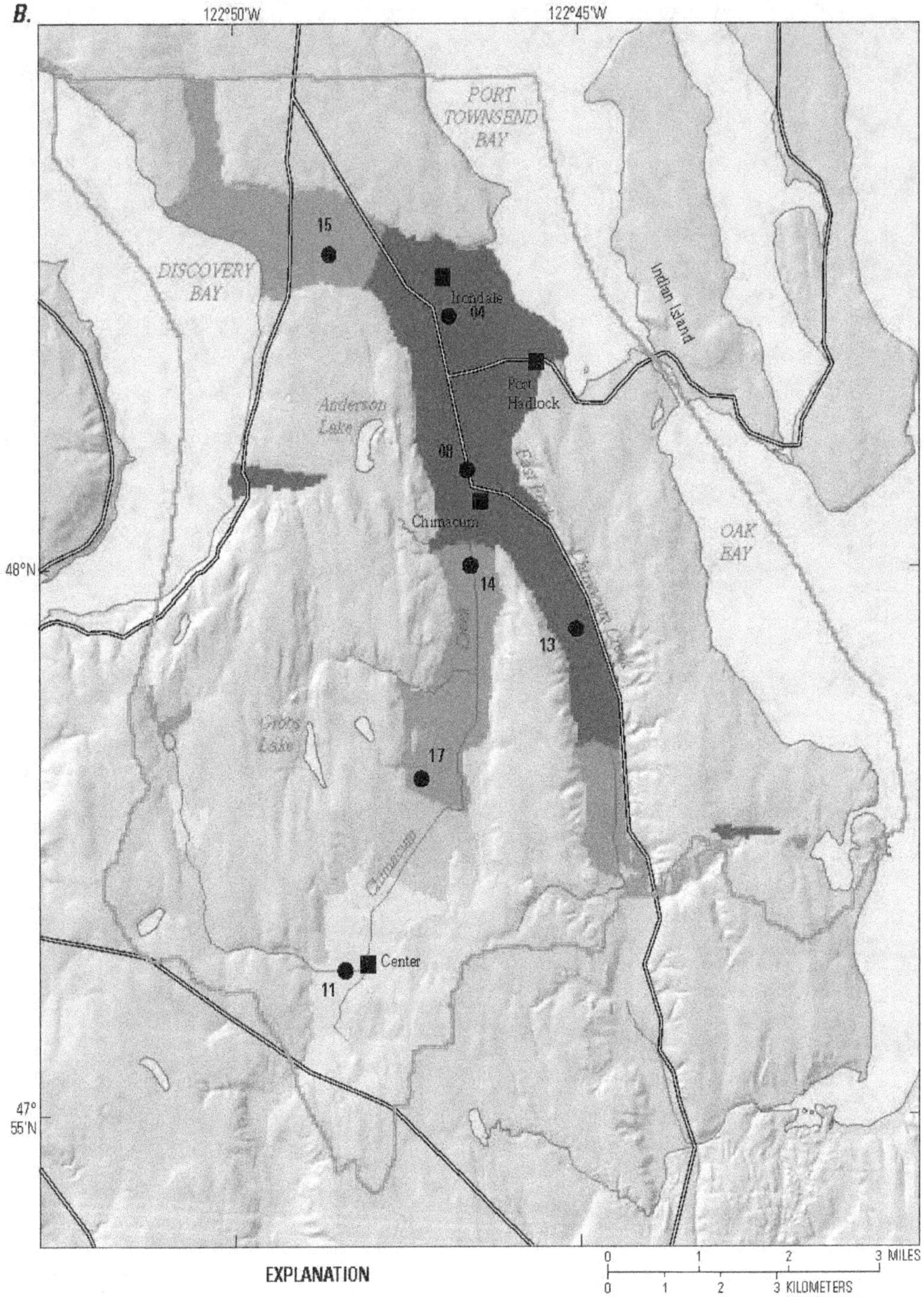

EXPLANATION

Horizontal hydraulic conductivity, Upper Aquifer unit (UA, model layer 2), in feet per day

0.4–1.0	50.1–100.0	Model boundary	
1.1–20.0	100.1–200.0	11	Pilot point and identifier
20.1–50.0	200.1–499.0		

Figure 8.—Continued

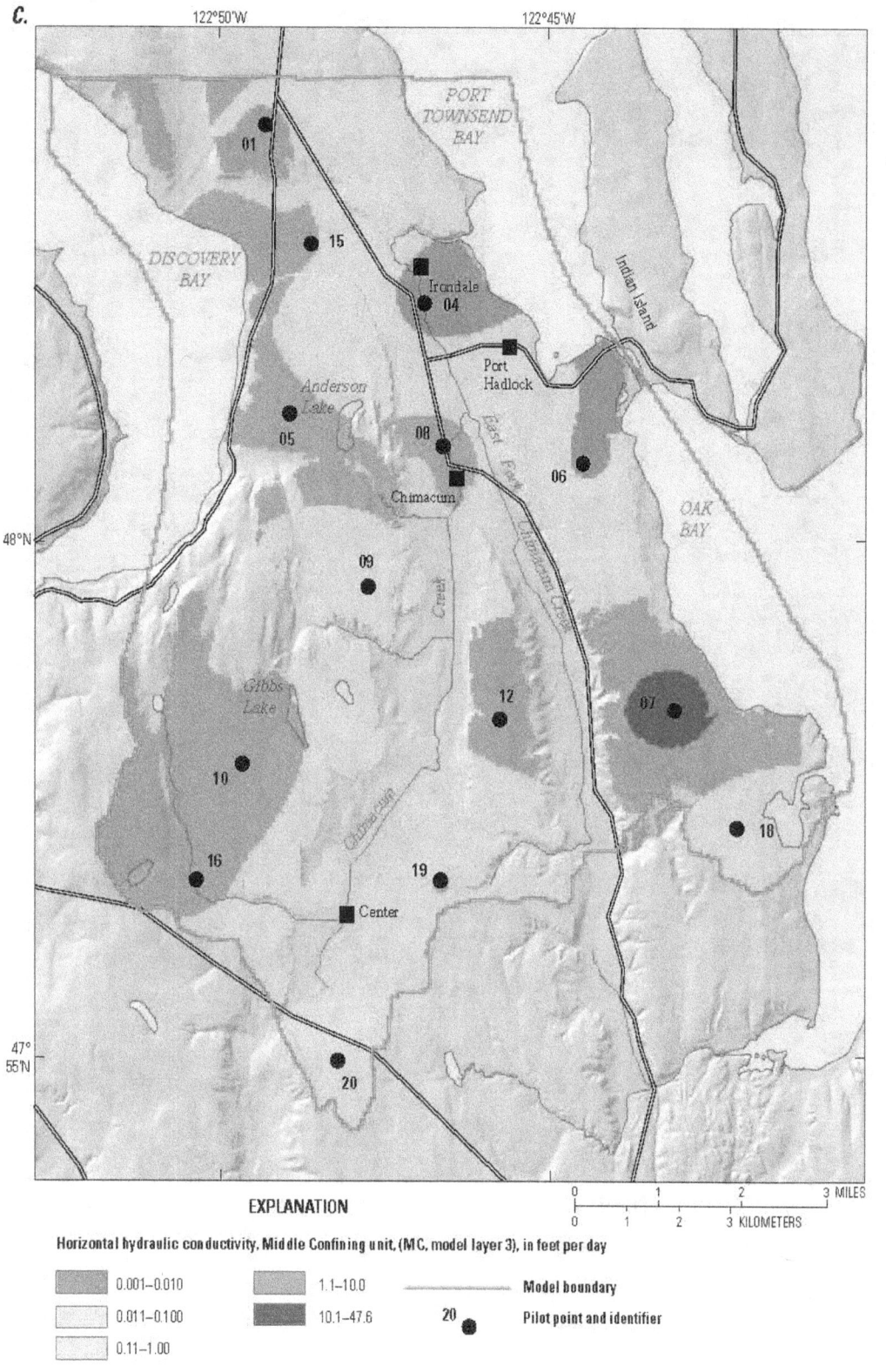

Figure 8.—Continued

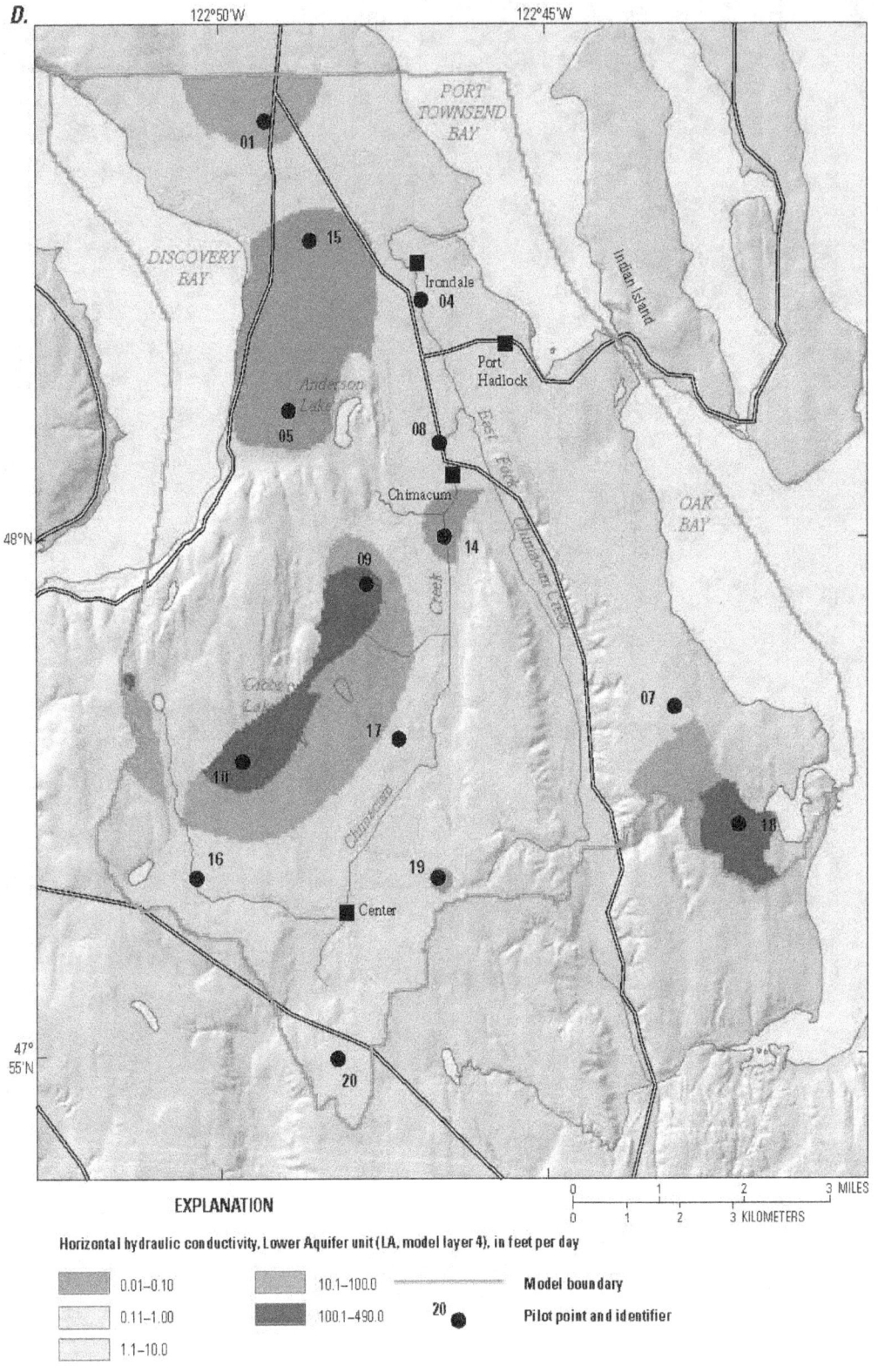

Figure 8.—Continued

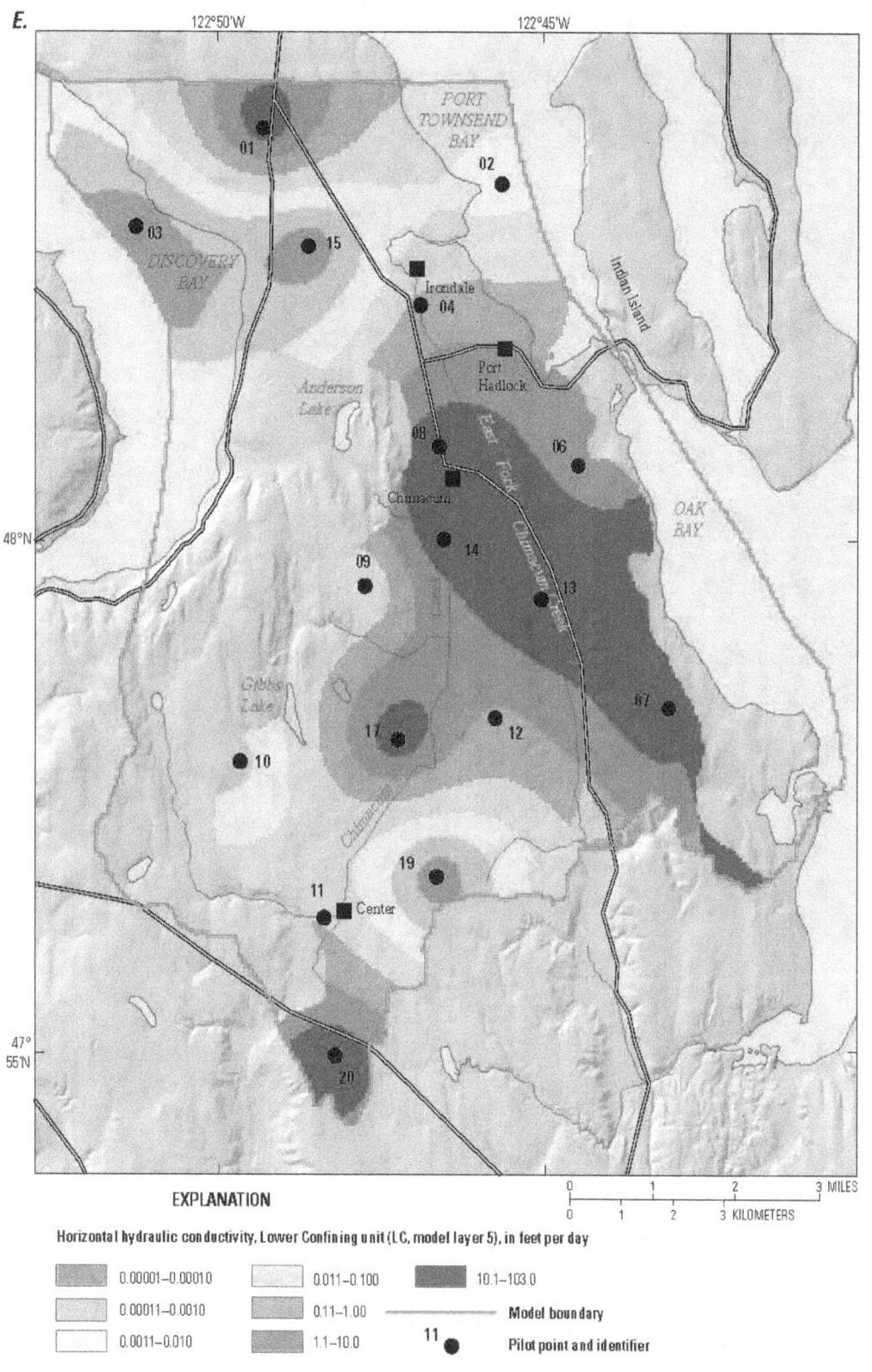

E.

EXPLANATION

Horizontal hydraulic conductivity, Lower Confining unit (LC, model layer 5), in feet per day

0.00001–0.00010	0.011–0.100	10.1–103.0
0.00011–0.0010	0.11–1.00	Model boundary
0.0011–0.010	1.1–10.0	11 ● Pilot point and identifier

Figure 8.—Continued

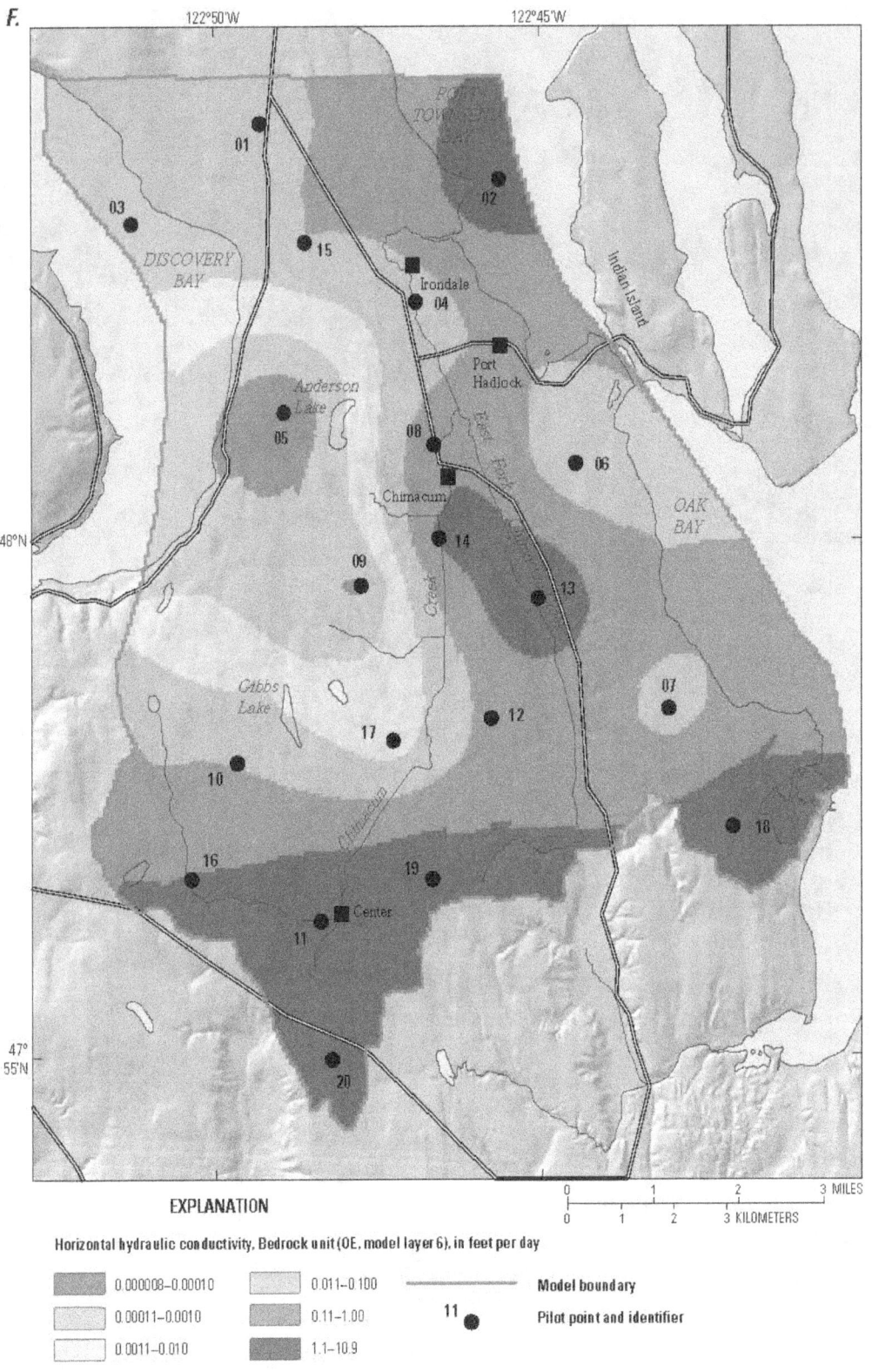

F.

EXPLANATION

Horizontal hydraulic conductivity, Bedrock unit (OE, model layer 6), in feet per day

0.000008–0.0010	0.011–0.100	Model boundary
0.00011–0.0010	0.11–1.00	Pilot point and identifier
0.0011–0.010	1.1–10.9	

Figure 8.—Continued

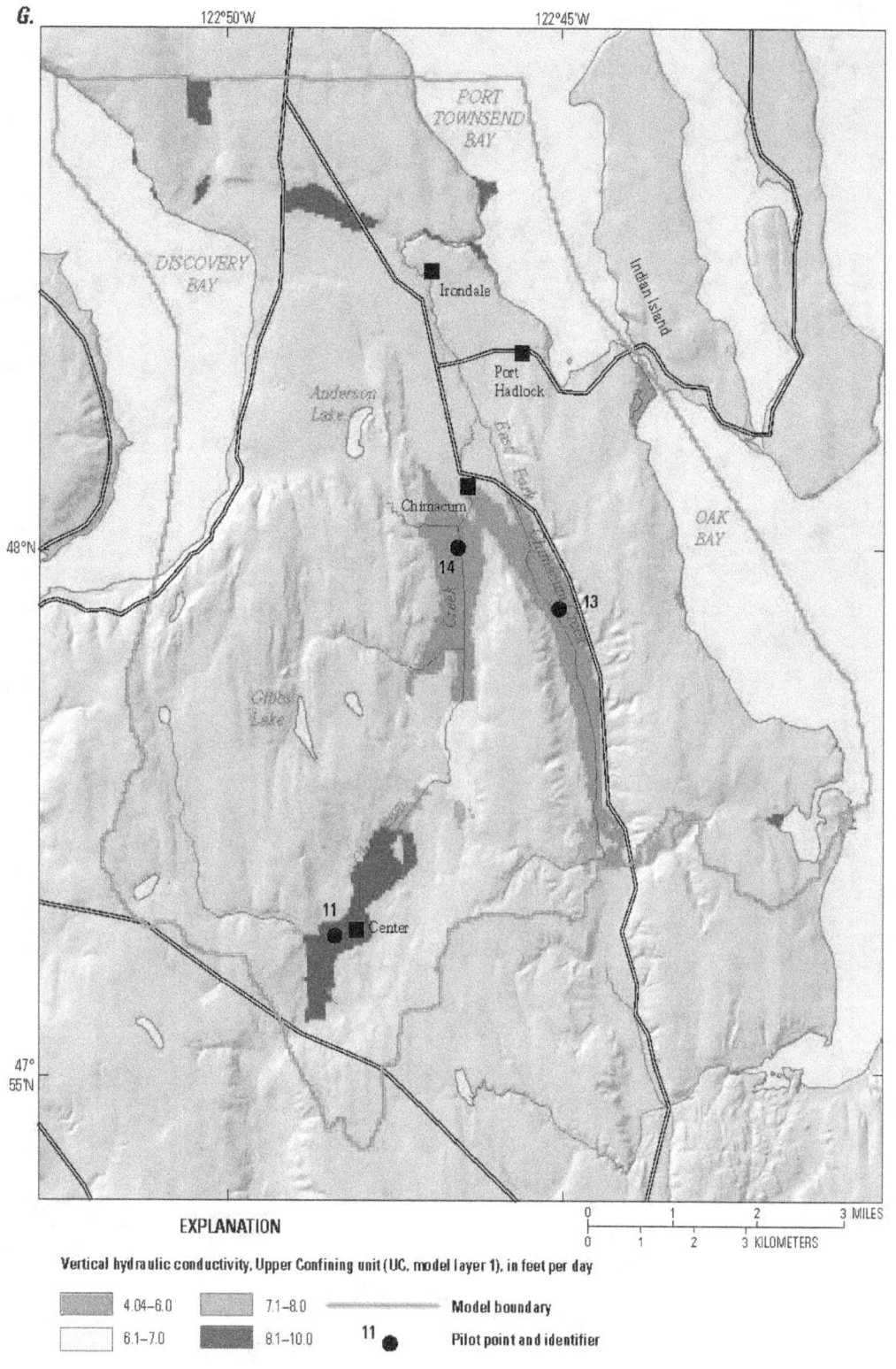

G.

EXPLANATION

Vertical hydraulic conductivity, Upper Confining unit (UC, model layer 1), in feet per day

4.04–6.0   7.1–8.0   Model boundary

6.1–7.0   8.1–10.0   **11** ● Pilot point and identifier

Figure 8.—Continued

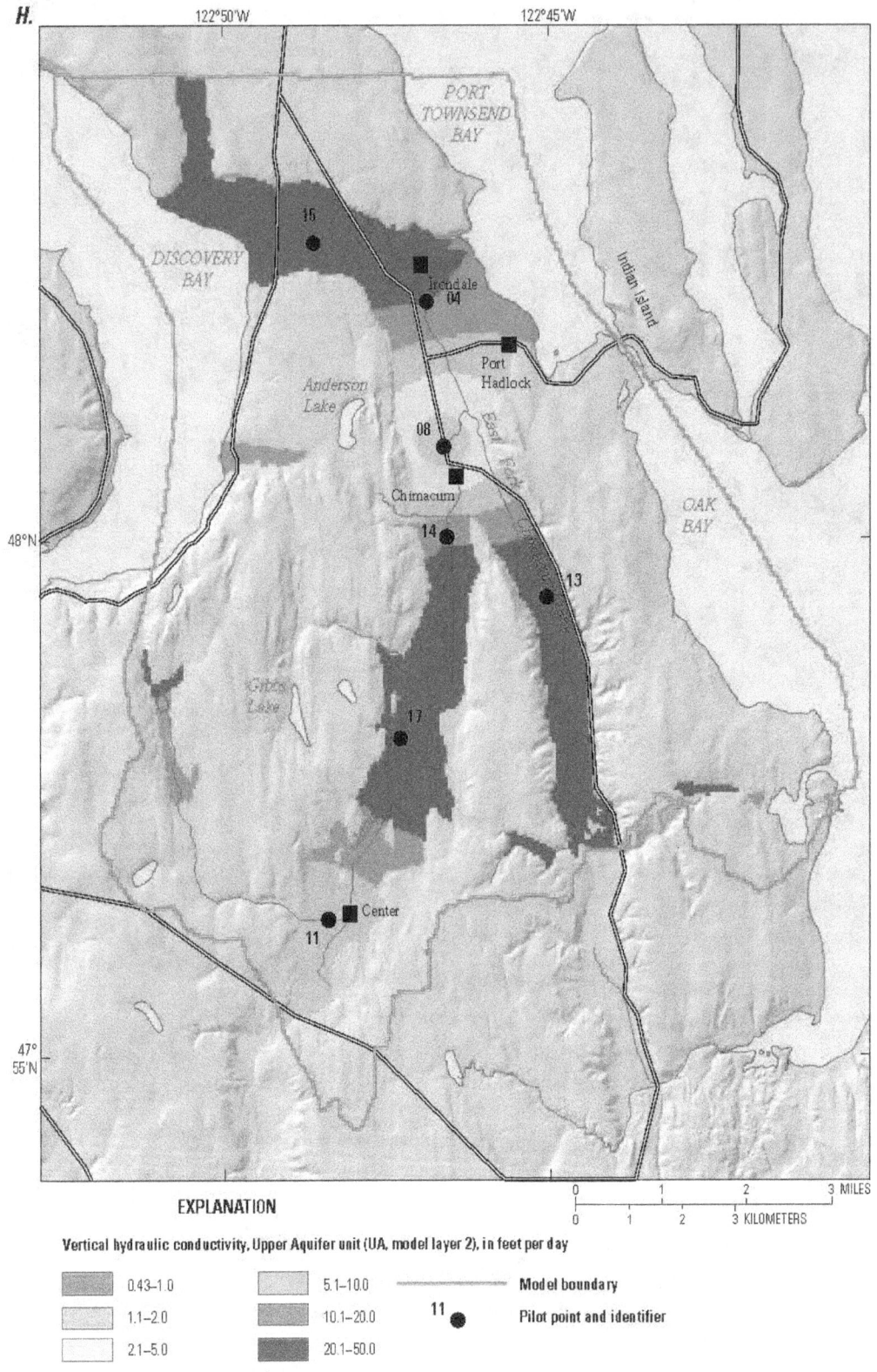

Figure 8.—Continued

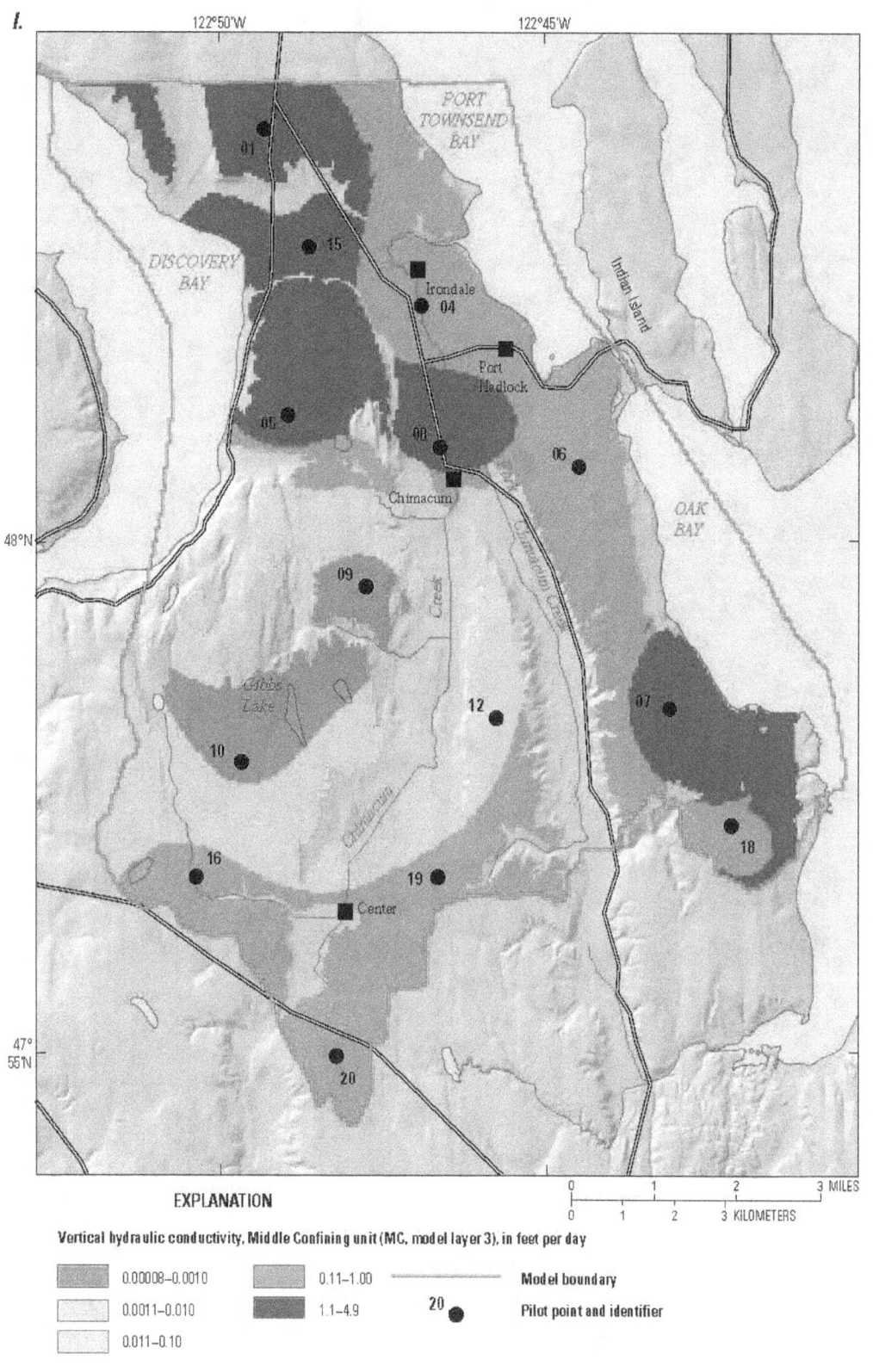

I.

EXPLANATION

Vertical hydraulic conductivity, Middle Confining unit (MC, model layer 3), in feet per day

0.00008–0.0010	0.11–1.00	—— Model boundary
0.0011–0.010	1.1–4.9	20 ● Pilot point and identifier
0.011–0.10		

Figure 8.—Continued

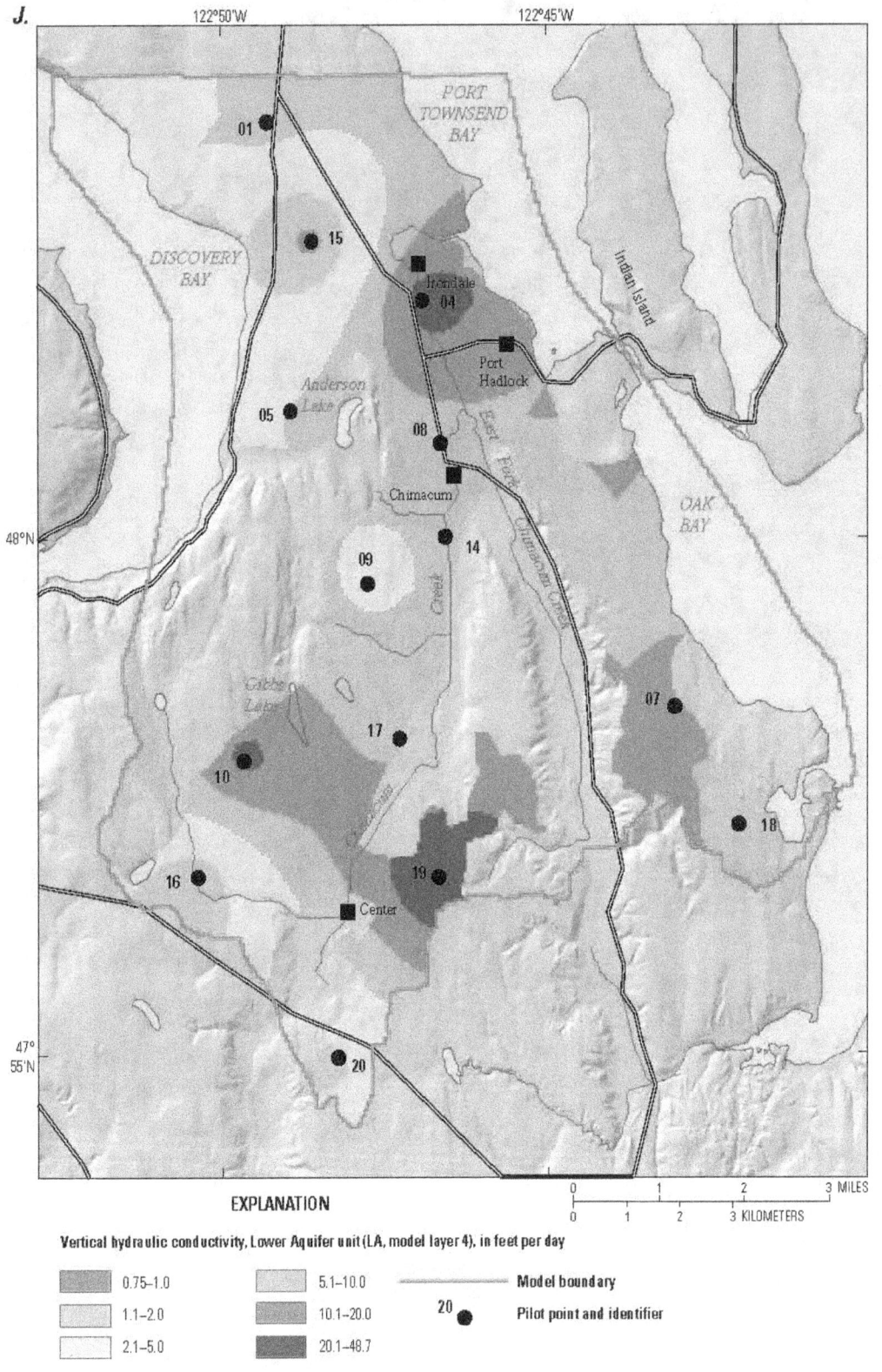

J.

EXPLANATION

Vertical hydraulic conductivity, Lower Aquifer unit (LA, model layer 4), in feet per day

0.75–1.0		5.1–10.0	Model boundary
1.1–2.0		10.1–20.0	20 ● Pilot point and identifier
2.1–5.0		20.1–48.7	

Figure 8.—Continued

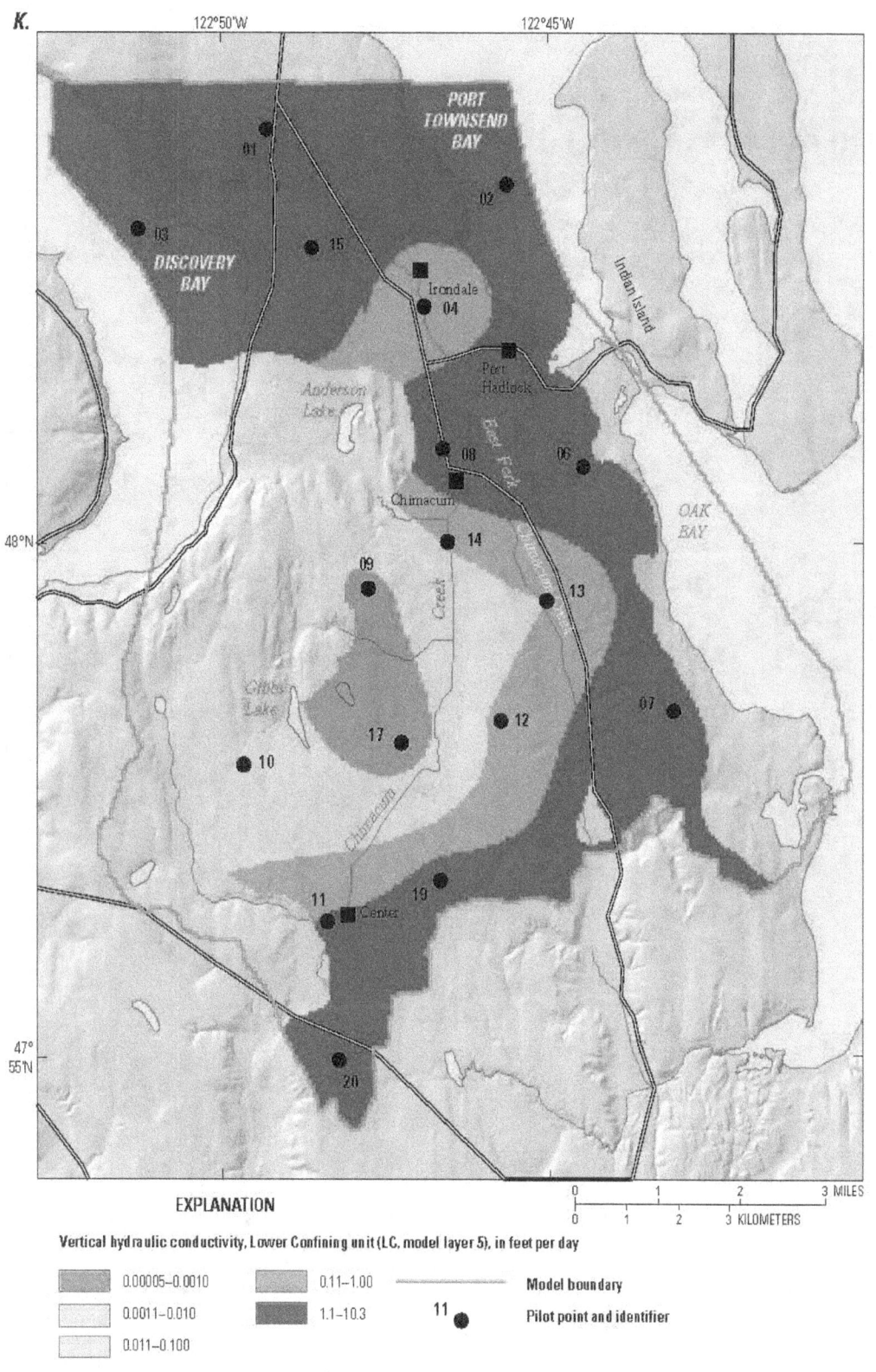

K.

EXPLANATION

Vertical hydraulic conductivity, Lower Confining unit (LC, model layer 5), in feet per day

0.00005–0.0010	0.11–1.00
0.0011–0.010	1.1–10.3
0.011–0.100	

Model boundary

11 ● Pilot point and identifier

Figure 8.—Continued

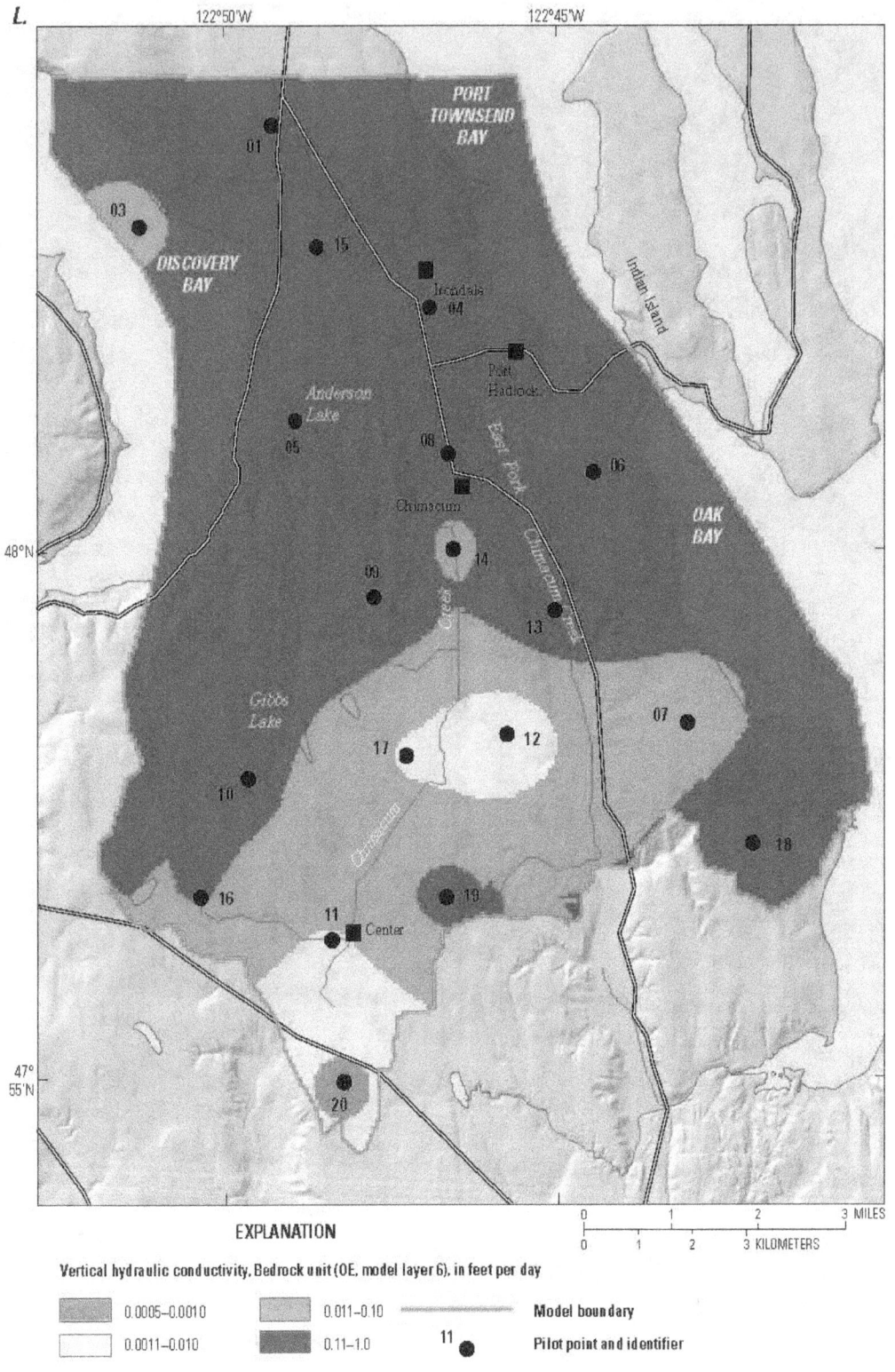

Figure 8.—Continued

Table 4. Water levels used for steady-state model calibration, Chimacum Creek Basin and vicinity, Jefferson County, Washington.

[NAVD 88, North American Vertical Datum of 1988]

Local well No.	Site identifier	Model layer	Number of measurements	Measured groundwater altitude (feet, NAVD 88)	Calibrated steady-state water level (feet, NAVD 88)	Steady-state residual (feet)
28N/01W-01C01	475712122445901	5	1	265.77	255.34	10.43
28N/01W-02A03	475714122452401	5	2	273.17	268.34	4.83
28N/01W-03H01	475653122464101	5	1	279.16	273.72	5.44
28N/01W-03J01	475640122464501	5	1	286.21	277.80	8.41
28N/01W-04R01	475629122480701	2	15	205.56	200.67	4.89
28N/01W-05A01	475708122491501	4	1	284.35	273.87	10.48
28N/01W-06H01	475656122504501	4	2	385.29	382.42	2.87
28N/01W-06J01	475646122503801	4	2	377.69	377.14	0.55
28N/01W-10B01	475622122470101	5	2	294.07	298.74	-4.67
28N/01W-10E01	475608122474201	5	2	245.74	258.42	-12.68
29N/01E-07M05	480106122440001	3	2	55.93	66.32	-10.39
29N/01E-19G02	475933122432101	5	1	93.74	103.07	-9.33
29N/01E-19P01	475908122433901	5	2	148.85	124.25	24.60
29N/01E-28N04	475816122412301	3	2	5.46	21.17	-15.71
29N/01E-29D01	475852122424201	3	1	92.61	89.99	2.62
29N/01E-29D02	475851122424101	6	1	32.81	66.41	-33.60
29N/01E-29D03	475852122424301	3	2	108.55	91.97	16.58
29N/01E-29D04	475851122424001	6	1	68.52	65.93	2.59
29N/01E-29D07	475851122424201	6	2	100.91	68.30	32.61
29N/01E-29R01	475815122413201	3	2	17.79	28.84	-11.05
29N/01E-32R02	475728122414601	2	2	45.09	43.29	1.80
29N/01E-33E03	475752122411601	3	1	39.45	31.95	7.50
29N/01E-33M05	475741122411101	3	2	30.45	28.41	2.04
29N/01W-03G02	480211122471301	4	2	75.68	52.91	22.77
29N/01W-03R01	480145122464201	4	31	73.76	75.74	-1.98
29N/01W-09J01	480105122481401	4	1	287.43	287.26	0.17
29N/01W-10A01	480129122464701	4	2	86.94	86.92	0.02
29N/01W-11L01	480115122455701	4	2	105.36	96.75	8.61
29N/01W-11Q01	480101122455101	4	2	92.29	105.39	-13.10
29N/01W-12H01	480118122440901	6	1	80.67	66.61	14.06
29N/01W-13M01	480016122451901	5	32	119.92	117.65	2.27
29N/01W-15B01	480047122465801	4	34	108.23	106.32	1.91
29N/01W-15R01	480007122465301	4	34	127.85	124.27	3.58
29N/01W-21E01	475942122490101	6	2	624.65	620.90	3.75
29N/01W-21E02	475938122490101	6	2	617.36	620.02	-2.66
29N/01W-21J01	475925122475201	5	1	244.47	207.12	37.35
29N/01W-22F03	475935122472601	5	2	183.13	175.08	8.05
29N/01W-23F01	475945122460201	5	19	136.68	120.87	15.81
29N/01W-24K03	475926122444101	5	35	132.33	124.18	8.15
29N/01W-26M03	475830122462101	2	1	122.79	132.39	-9.60
29N/01W-27C01	475856122471801	4	2	221.27	206.59	14.68
29N/01W-27E01	475851122474401	5	1	173.8	179.48	-5.68
29N/01W-27F01	475844122472501	5	2	177.21	171.08	6.13
29N/01W-31B02	475802122510101	4	2	442.95	445.00	-2.05
29N/01W-33R02	475720122480401	4	2	248.26	254.57	-6.31
29N/01W-34C01	475801122472601	4	2	185.61	204.76	-19.15
29N/01W-34F01	475758122473001	4	2	242.07	213.03	29.04
29N/01W-35J01	475734122453701	5	32	276.41	270.01	6.40
29N/01W-35L01	475734122461101	5	1	276.76	262.80	13.96
29N/01W-36A01	475807122441701	5	1	150.41	138.65	11.76

Table 4. Water levels used for steady-state model calibration, Chimacum Creek Basin and vicinity, Jefferson County, Washington.— Continued

[NAVD 88, North American Vertical Datum of 1988]

Local well No.	Site identifier	Model layer	Number of measurements	Measured groundwater altitude (feet, NAVD 88)	Calibrated steady-state water level (feet, NAVD 88)	Steady-state residual (feet)
30N/01W-28F02	480358122483501	4	2	165.56	112.48	53.08
30N/01W-28M02	480350122491001	4	2	124.3	116.62	7.68
30N/01W-29A01	480411122491501	4	1	141.85	120.45	21.40
30N/01W-32G01	480305122495201	4	1	2.69	36.38	-33.69
30N/01W-33N02	480245122490201	4	1	38.95	23.70	15.25
30N/01W-34E01	480308122472801	4	1	78.2	38.51	39.69
30N/01W-34H01	480306122464201	4	2	8.2	24.66	-16.46

Table 5. Surface-water discharge measurements used for steady-state model calibration, Chimacum Creek Basin and vicinity, Jefferson County, Washington.

[Data from Simonds and others (2004). **Map identifier:** Locations shown in figure 7. Values in cubic feet per second]

Map identifier	Site location	Number of measurements	Average flow measurement	Calibrated steady-state flow	Steady-state residual
CS1	Chimacum Creek, 20 feet upstream of sediment basin, and 0.8 mile west of Center	3	0.65	0.12	0.52
CS3	Chimacum Creek, 50 feet downstream of West Valley Road, and 0.6 mile northwest of Center	3	2.50	2.52	-0.02
CS4	Chimacum Creek, at Center Road bridge, and 1.7 miles north of Center	3	2.98	3.79	-0.81
CS5	Chimacum Creek, 100 feet downstream of road bridge, and 2.4 miles north of Center	3	4.08	5.15	-1.07
CS6	Naylor Creek, 10 feet upstream from weir, 50 feet downstream of West Valley Road, and 2.8 miles north of Center	3	0.24	0.00	0.24
CS7	Putaansuu Creek, 10 feet downstream of West Valley Road, and 0.9 mile southwest of Chimacum	3	0.12	0.03	0.09
CS8	Chimacum Creek, at Rhody Drive bridge, and 0.3 mile west of Chimacum	3	3.70	4.73	-1.03
CS9	East Fork Chimacum Creek, 30 feet upstream of Egg and I Road, and 2.0 miles north of Beaver Valley	3	0.67	0.03	0.64
CS10	East Fork Chimacum Creek, upstream of culvert, and 3.2 miles south of Chimacum	3	1.59	0.15	1.44
CS11f	East Fork Chimacum Creek, Beaver Valley Road, and 0.3 mile southeast of Chimacum	2	1.24	0.00	1.24
CS12	East Fork Chimacum Creek, 20 feet downstream of Chimacum Road, and 0.6 mile north of Chimacum	3	1.21	0.08	1.13
CS13	Chimacum Creek, at PUD gage, 50 feet upstream of footbridge, 300 feet east of end of Hilda Road, 1.2 miles north of Chimacum, and at mile 2.3	3	7.14	2.27	4.87
CS14	Chimacum Creek, 0.7 mile upstream of mouth	84	11.29	7.41	3.88

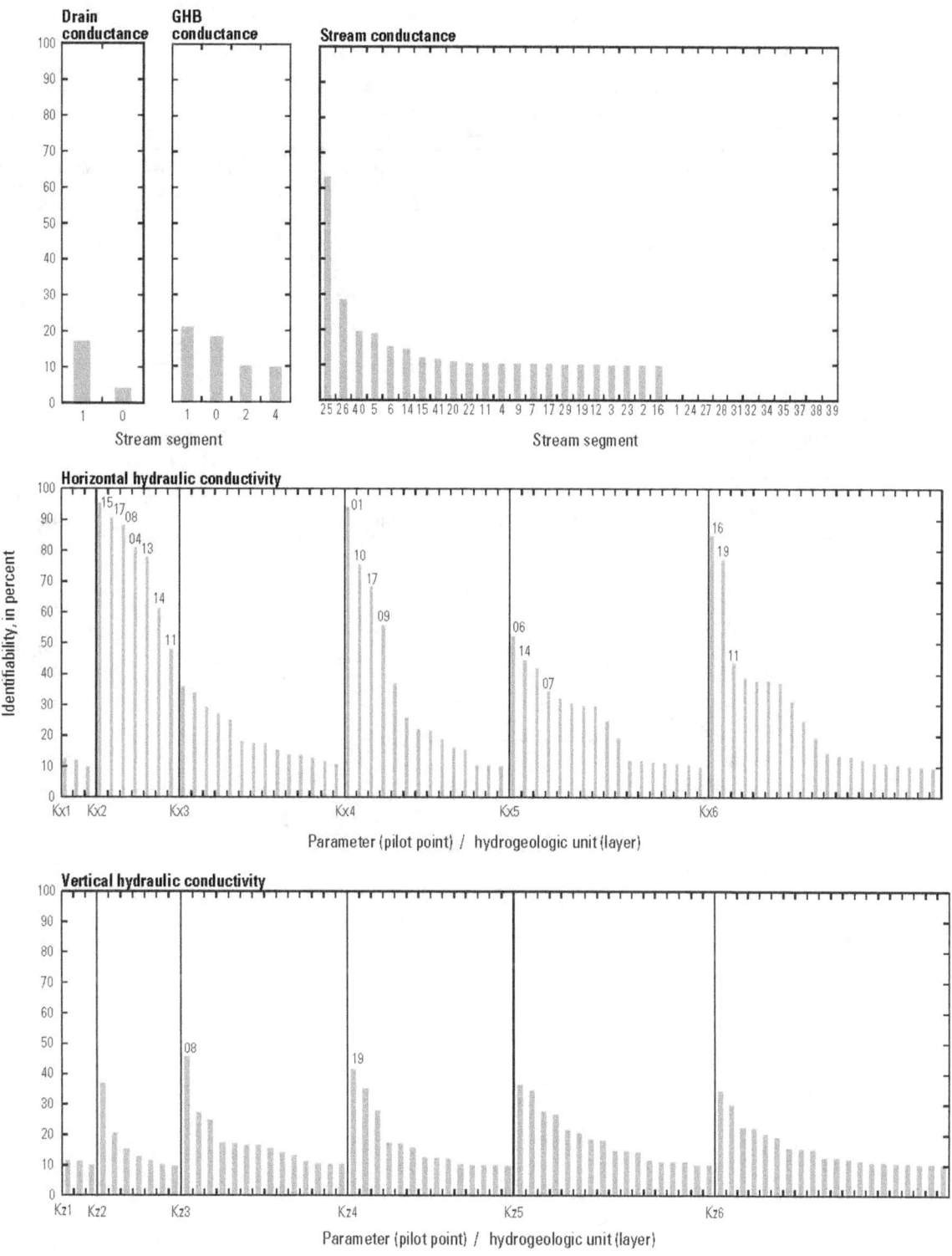

Figure 9. Identifiability plots for parameters (selected pilot points labeled) used in model calibration, Chimacum Creek Basin and vicinity, Jefferson County, Washington. Stream segments are shown in figure 3.

Table 6.   Final values for model calibration parameters, Chimacum Creek Basin and vicinity, Jefferson County, Washington.

[**Abbreviations:** <, less than; NA, not applicable]

	Horizontal hydraulic conductivity (Kh) (feet per day)					
Model layer	Number of active cells	Geometric mean	Minimum	Maximum	Change from initial estimate	Median thickness (feet)
1	3,136	0.375	0.002	32.04	× 0.38	30
2	9,150	167.6	0.436	498.8	× 10.1	<5
3	25,035	0.396	0.001	47.62	× 0.49	51
4	21,795	1.627	0.01	489.6	× 0.48	62
5	28,966	0.152	$1.0(10)^{-5}$	103.0	× 0.06	274
6	44,775	0.08	$7.0(10)^{-6}$	10.9	× 0.42	1,476

	Vertical hydraulic conductivity (Kv) (feet per day)			
Model layer	Geometric mean	Minimum	Maximum	Change from initial estimate
1	6.27	4.04	10	× 1.13
2	17.9	0.427	50	× 0.69
3	0.184	$7.6(10)^{-5}$	4.94	× 0.64
4	6.69	0.751	43.3	× 0.64
5	0.73	$5.5(10)^{-5}$	10.3	× 0.71
6	0.165	$4.9(10)^{-4}$	0.977	× 1.37

	Drain conductance (square feet per day)		
Drain group	Application	Final value	Change from initial estimate
0	Small streams, shallow bedrock	1,000	× 0.043
1	Coastal seeps	25,852	× 1.63

	General head boundary (GHB) conductance (square feet per day)		
GHB group	Application	Final value	Change from initial estimate
0	Model boundaries in Puget Sound	907.7	× 0.00087
1	Lakes	$5.34(10)^8$	× 28.6
2	North boundary in Unit 2	100	NA
4	North boundary in Unit 4	$1.53(10)^5$	NA

	Stream conductance (square feet per day)	
Stream reach	Final value	Change from initial estimate
1	3.2	× 0.298
2	66.0	× 0.207
3	763	× 77.2
4	96,612	× 764
5	0.37	× 0.0148
6	1,140	× 0.170
7	0.36	× 0.0006
9	0.87	× 0.0044
11	3,769	× 1.261
12	389	× 6.83
14	378,883	× 22.8
15	2,028	× 41.1
16	10.2	× 0.494
17	2.61	× 1.208
19	3,311	× 22.1
20	330	× 3.33
22	145	× 6.71
23	110.7	× 4.18
24	529	× 2.90
25	2,881	× 10.15
26	66,233	× 4.02
27	55.2	× 8.07
28	74.7	× 0.109
29	14.6	× 5.38
31	56,467	× 216
32	53.3	× 0.569
34	270,211	× 143.7
35	23,717	× 28.3
37	6,543	× 148.7
38	711	× 7.61
39	15,439	× 13.17
40	11,589	× 178.3
41	8.77	× 0.481

Assessment of Steady State Calibration

The results of the calibration were assessed by comparing measured and simulated groundwater levels and stream base flows, and by examining the mean and standard error of residuals (difference between measured and simulated values). The sign of the mean of residuals (bias) indicates whether the model is generally over- or under-predicting values (negative and positive mean of residuals, respectively). The standard error of residuals is a measure of how much variation there is in residual values greater than and less than the mean residual value. Another method of evaluating the fit of calibration is to assess the ratios of the standard error of model residuals to the range of observations. Ratios less than 0.10 generally represent a good model fit (Kuniansky and others, 2004).

Table 7 shows the steady-state calibration statistics for groundwater levels (by hydrogeologic unit) and for stream base flow gains or losses. The minimum standard error on the mean between simulated and measured groundwater levels (2.79 ft) occurred in layer 5 (hydrogeologic unit LC); the maximum standard error (8.89 ft) occurred in layer 6 (bedrock). Layers 2 and 3 (UA and MC) had the lowest absolute value of mean residuals, indicating that simulated groundwater levels in these units had the lowest model bias. The ratio between the standard error of the residuals and the water-level range was less than 5 percent for all model layers.

A plot of simulated and measured groundwater-level altitudes by hydrogeologic unit (fig. 10) indicates good agreement with some bias in layers 4 and 6.

Table 7. Calibration statistics by hydrogeologic unit and base flow, Chimacum Creek Basin and vicinity, Jefferson County, Washington.

[**Hydrogeologic unit:** UC, Upper Confining unit; UA, Upper Aquifer unit; MC, Middle Confining unit; LA, Lower Aquifer unit; LC, Lower Confining unit; OE, Bedrock unit. **Abbreviations:** ft, feet; ft^2, square feet]

Hydrogeologic unit (model layer)	Number of observations	Mean of residuals (ft)	Standard error of residuals (ft)	Mean of absolute values of residuals (ft)	Objective function (sum of squared residuals) (ft^2)	Range of observed values (ft)	Standard error of residual for range of observations (percent)
Heads in UC (1)	0						
Heads in UA (2)	3	-0.97	4.41	5.43	119	45.09–205.56	2.7
Heads in MC (3)	7	-1.20	4.39	9.41	819	5.46–108.55	4.3
Heads in LA (4)	23	6.04	4.00	14.11	8,921	2.69–442.95	0.9
Heads in LC (5)	18	7.29	2.79	10.89	3,332	93.74–294.07	1.4
Heads in OE (6)	6	2.79	8.89	14.88	2,418	32.81–624.65	1.5
Total of all heads	57	4.84	2.12	12.14	15,610	2.69–624.65	0.3
Base flow observations	15	[1]0.61	[1]0.42	[1]1.16	[2]42.88	[1]0.05–11.29	3.7

[1] Value in cubic feet per second (ft^3/s).

[2] Value in cubic feet per second squared [(ft^3/s)^2].

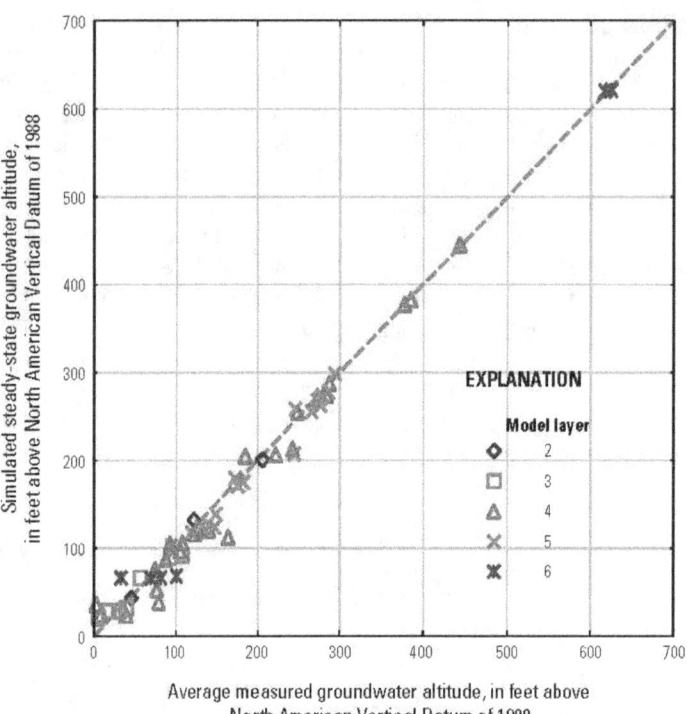

Figure 10.   Simulated and measured groundwater-level altitude for the calibrated model for steady-state conditions, Chimacum Creek Basin and vicinity, Jefferson County, Washington.

The simulated water-level altitudes reasonably simulate the measured magnitudes and general groundwater-head patterns described in Jones and others (2011). Simulated hydraulic heads for layers UA, LA, and LC are shown in figure 11. The residuals are the differences between the measured target value minus model simulated value of heads, so positive residuals are at sites where the model simulation predicts heads that are lower than measured (under prediction), and negative residuals are at sites where the simulation predicts heads that are higher than measured (over prediction). The spatial distribution of the hydraulic-head residuals does not indicate any major patterns of bias.

Simulated steady-state groundwater-level altitudes in the UA unit indicate flow generally moving down the valleys to the north from the drainage divide. Simulated heads were generally within 10 ft of the measured heads, which indicates a good model fit.

Simulated steady-state groundwater-level altitudes in the LA unit indicate flow generally moving down valley from the drainage divide to discharge in Discovery, Port Townsend, and Oak Bays. A groundwater divide occurs midway between Discovery Bay and Port Townsend Bay. Most simulated heads were within 10 ft of the measured heads, which indicates a good fit.

Simulated steady-state groundwater-level altitudes in the LC unit indicate flow generally moving down valley from the drainage divide to discharge in Discovery, Port Townsend, and Oak Bays. A groundwater divide occurs midway between Discovery Bay and Port Townsend Bay. Simulated head residuals in unit LC were larger than in units UA and LA, which reflects the heterogeneity of the unit.

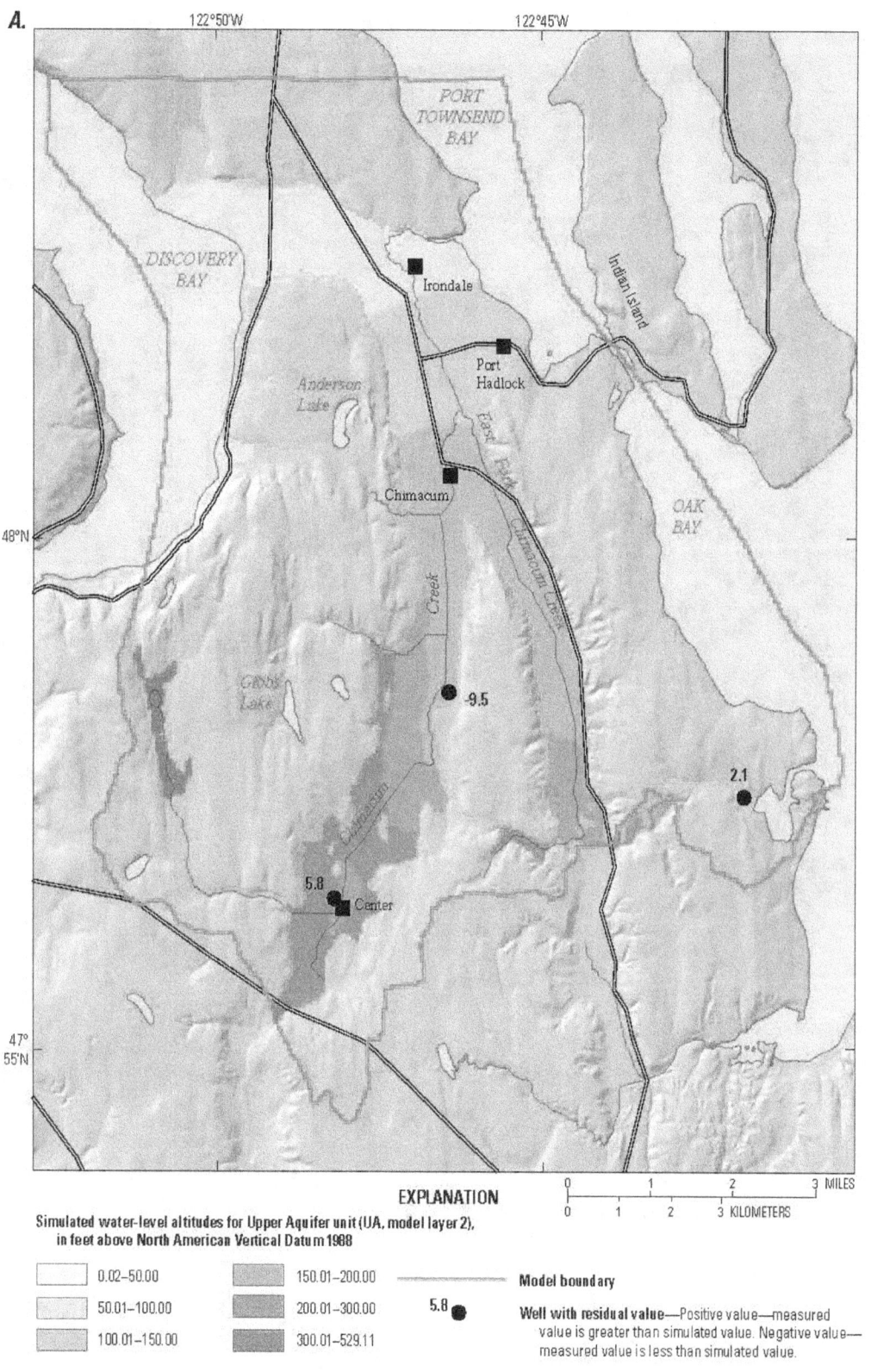

A.

Figure 11. Simulated water-level altitudes and residuals for the calibrated model for steady-state conditions, model layers (*A*) 2 (unit UA), (*B*) 4 (unit LA), and (*C*) 5 (unit LC), Chimacum Creek Basin and vicinity, Jefferson County, Washington.

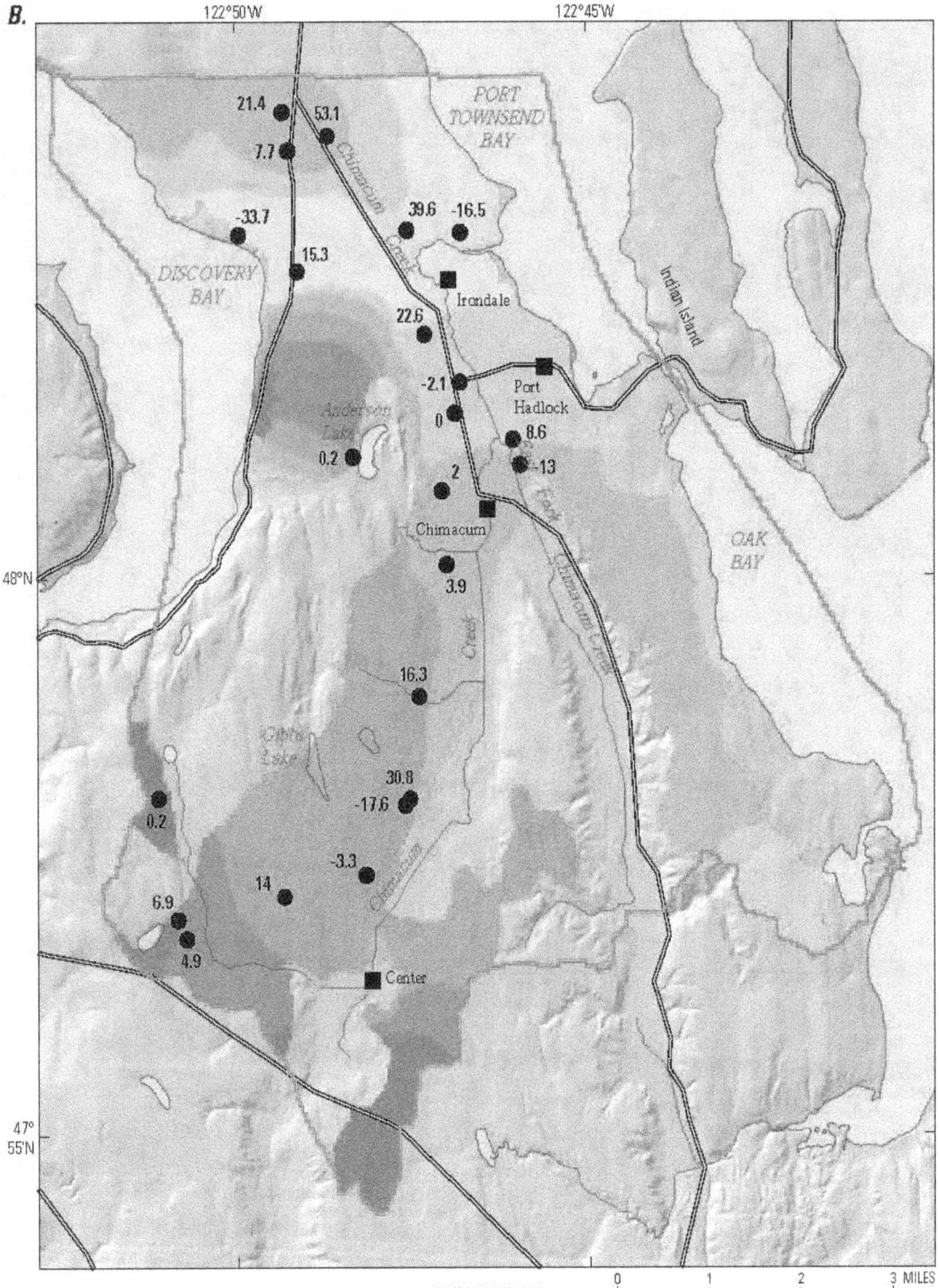

B.

EXPLANATION

Simulated water-level altitudes for Lower Aquifer unit (LA, model layer 4),
in feet above North American Vertical Datum 1988

-21.77–50.00	200.01–300.00	Model boundary
50.01–100.00	300.01–400.00	4.9 ● Well with residual value—Positive value—measured
100.01–200.00	400.01–824.24	value is greater than simulated value. Negative value— measured value is less than simulated value.

0   1   2   3 MILES

0   1   2   3 KILOMETERS

Figure 11.—Continued

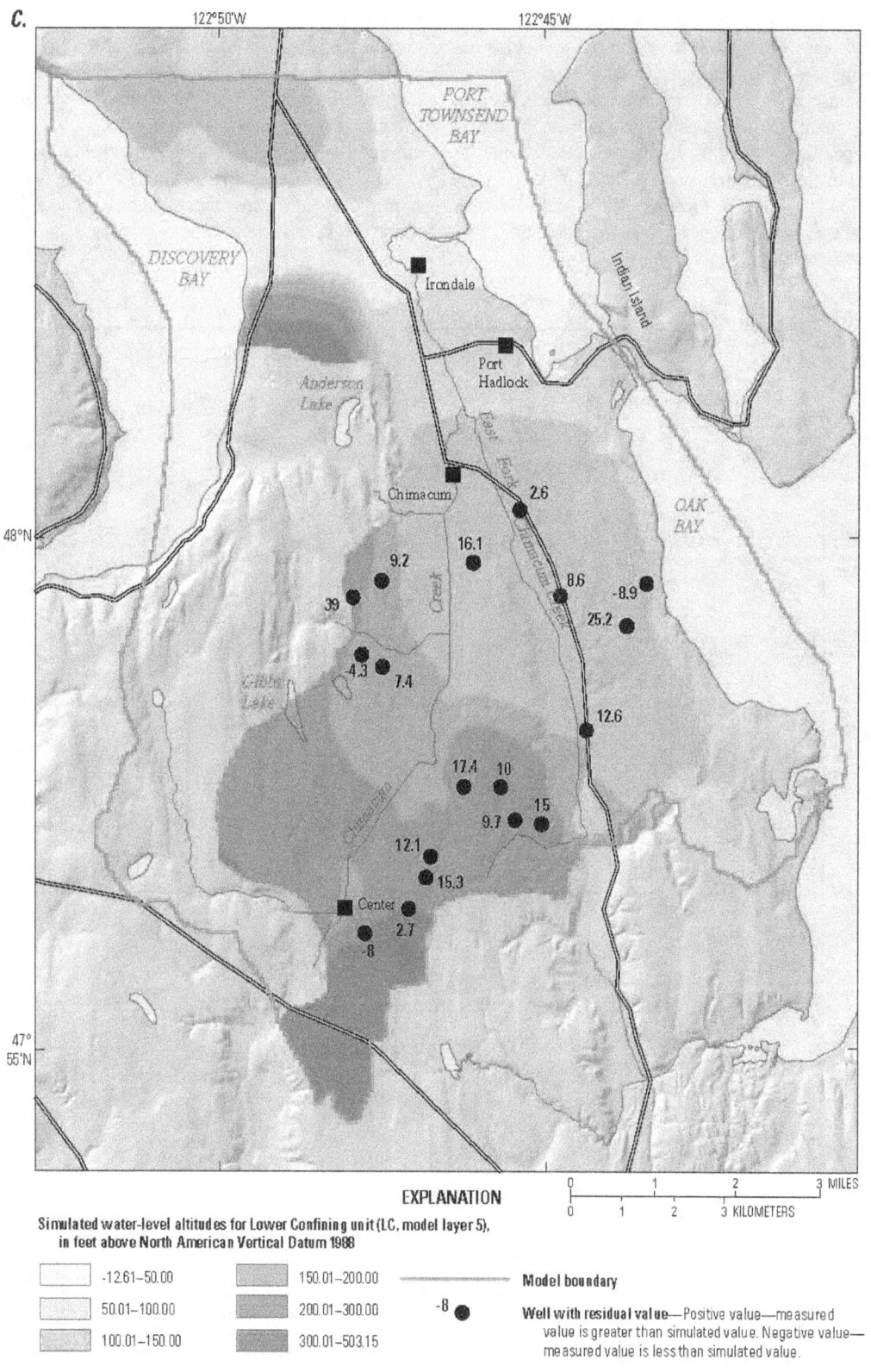

EXPLANATION

Simulated water-level altitudes for Lower Confining unit (LC, model layer 5), in feet above North American Vertical Datum 1988

-12.61–50.00	150.01–200.00
50.01–100.00	200.01–300.00
100.01–150.00	300.01–503.15

Model boundary

-8 ● Well with residual value—Positive value—measured value is greater than simulated value. Negative value—measured value is less than simulated value.

Figure 11.—Continued

The average measured and simulated steady-state groundwater discharge to streams (base flow) were compared (fig. 12). Synoptic base flow measurements were made throughout Chimacum Creek, and mean monthly base flow at the streamgage on Chimacum Creek was calculated using hydrograph separation. Base flow discharge is reasonably well simulated by the calibrated model except for the East Fork Chimacum Creek. On average, the model simulation underpredicted the amount of base flow in the East Fork of Chimacum Creek and at two locations downstream of the confluence between Chimacum Creek and the East Fork (CS13 and CS14). This is likely due to the humic bogs present along the East Fork (locally known as "Magical Dirt"), which drain slowly and give the recession curve a logarithmic shape similar to one affected by large amounts of bank storage. This is evidenced by the dark brown color of the water in the reach during the summer months as the humic bogs slowly discharge to the creek.

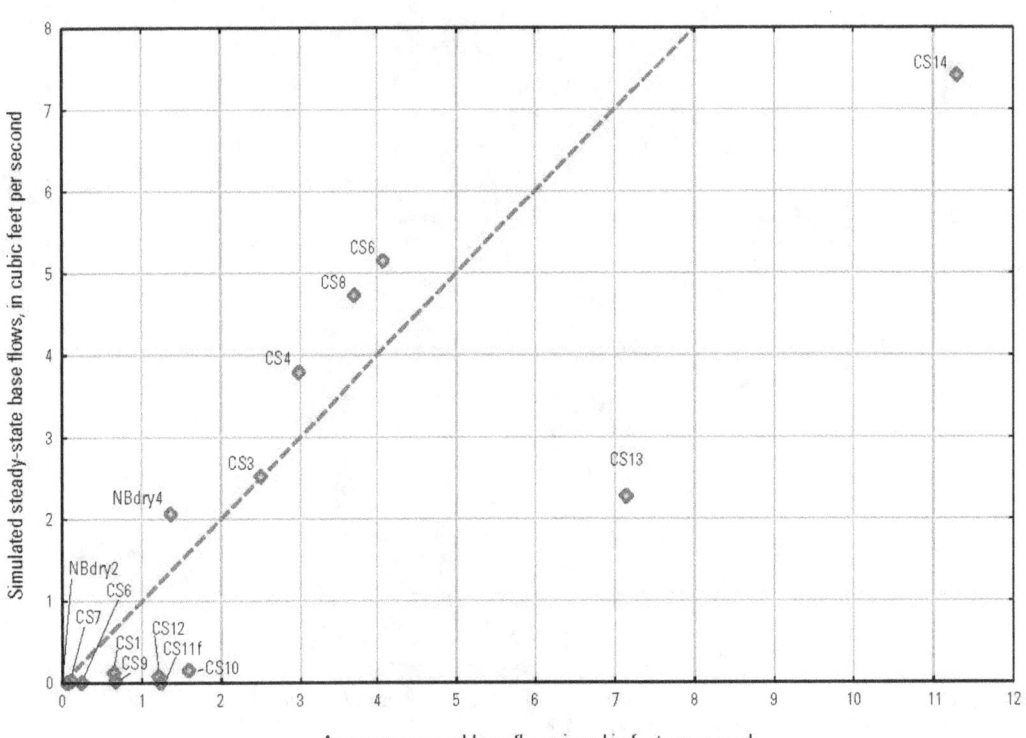

Figure 12. Simulated steady-state base flows and measured base flows, Chimacum Creek Basin and vicinity, Jefferson County, Washington.

Model Limitations

A groundwater-flow model represents a complex, natural system with a set of mathematical equations that describe the groundwater-flow system. Intrinsic to the model is the error and uncertainty associated with the assumptions that are made. Hydrologic-modeling errors typically are the consequence of a combination of input data, representation of the physical processes by the algorithms of the model, and parameter estimation during the calibration procedure (Troutman, 1985). Examples of the three types of model errors, and how those errors limit application of the model, are:

1. Input data on types and thicknesses of hydrogeologic units, water levels, surface-water diversions, and hydraulic properties represent only approximations of actual values. Model-discretization errors (including effects of averaging elevation information over the model cell size) result from inaccuracies in the geometric representation of hydrogeologic units, and in the representation of the bedrock areas and their contact with unconsolidated units.

2. All the physical processes in a watershed are not represented completely or "captured" in a numerical model. The simplifying assumptions and generalizations that are incorporated into a model affect the results of the simulation, but determining whether a weakness in a simulation is attributable to input data error or model shortcomings is difficult.

3. Errors in parameter estimates occur when improper values are selected during the calibration process. Various combinations of parameter values can result in low residual error, yet improperly represent the actual system. An acceptable degree of agreement between simulated and measured values does not guarantee that the estimated model-parameter values uniquely and reasonably represent the actual parameter values. The use of nonlinear regression and associated statistics, such as composite scaled sensitivities and correlation coefficients, removes some of the effects of non-uniqueness, but not entirely.

If the regional groundwater-flow model is used appropriately, the effects of the simplifications and other potential errors can be limited. If the model is used for simulations beyond which it was designed, however, the generalizations and assumptions used could significantly affect the results. For example, although the model cell size is 200 ft, the model simulation results should not be used as a basis for decisions at that scale, such as "What would happen if I put a well here versus 100 feet south?"

With respect to the treatment of parts of the upper three models layers as confined, though conceptually they are unconfined, the following excerpts from Faunt and others (2011) provide some background.

"The MODFLOW model does not use MODFLOW's unconfined model layer capability. Instead the DVRFS unconfined aquifer, in which the actual saturated thickness varies over time is approximated using a model layer in which the simulated saturated thickness used to calculate transmissivity remains constant over time. Within the constant-saturated-thickness layer, storage changes over a given period of time are calculated as specific yield times simulated head change. This approximation greatly enhances computational speed and stability, especially during calibration when some attempted sets of parameter values produce computed heads that differ substantially from measured heads."

"Unconfined systems can be difficult to model because of extra computational burden and possible nonlinear instability. But if drawdowns are relatively small compared to initial saturated thickness of the unconfined unit(s), speed and stability can be gained with minimal loss of accuracy by holding the saturated thickness constant. Approximating an unconfined aquifer as having constant saturated thickness is a well-known modeling technique…"

"For simulation of steady-state flow, one can iterate, setting the top of the model to the estimated top of the unconfined system (the water-table elevation), solving for new heads, resetting the model top to the new simulated water table, etc., until the model coincides with the water table."

"For simulation of transient flow [or steady-state simulations of conditions different from calibrated steady-state conditions—transient simulations with one infinite time step] one can set the model top to the initial water-table elevation and proceed to simulate drawdowns with the constant-saturated-thickness model. The simulated drawdowns are accurate as long as they remain a modest fraction of the initial saturated thickness, because the transmissivity (thickness multiplied by hydraulic conductivity) of the constant-saturated-thickness system will remain close to that of the unconfined system." "Our experience suggests that satisfactory accuracy can be expected for drawdowns of less than 10 % of initial saturated thickness, and that even larger relative drawdowns can yield acceptable results."

The iterative process to reduce the modeled saturated thickness (h) from the thickness of the entire model layer to something less, reflecting model-simulated saturated thickness, during calibration of the Chimacum Creek model, was not conducted because it was felt that the calibration of conductivity values (K) using constant h was sufficient given the anticipated stresses to be imposed in future water-use scenarios. A reduction in modeled saturated thickness would correspondingly reduce the absolute magnitude of acceptable head changes (10 percent of a smaller h) due to changes in system stresses. Modeled responses to increased pumpage (scenarios "Probable Future Use" and "Full Beneficial Use") or decreased return flow ("Sanitary Sewer") were lower heads on the order of fractions of a foot, suggesting model results should be within an acceptable margin of error. However, future applications of the Chimacum Creek model should be made with the awareness that changes in water use or recharge that cause much larger reductions in simulated head may be approaching the limits of acceptable fractions of the saturated thickness for which the assumption of constant saturated thickness and thereby constant transmissivity is valid.

Model Applications

The calibrated model was used to derive components of the groundwater budget and to estimate the steady-state response of the regional system to new stresses, such as increased groundwater withdrawals. Water-resource managers can use this information to make informed decisions when planning for future groundwater development.

Model-Derived Groundwater Budget

A groundwater budget for average conditions during the calibration period (October 1, 1994–September 30, 2009) in the terrestrial part of the model area is expressed by the following equation:

$$GWin + R = GWout + D + \Delta S \qquad (1)$$

where

$GWin$ is groundwater inflow to the model area,

$GWout$ is groundwater outflow from the model area,

R is recharge,

D is discharge, and

ΔS is change in groundwater storage.

Recharge to the groundwater system occurs primarily as precipitation and seepage from streams and lakes. Return flow occurs as seepage from septic systems, and deep percolation of irrigation water. Discharge from the groundwater system

occurs as seepage to streams and lakes, as evaporation of groundwater from soils and transpiration from plants, as submarine seepage to inlets of Puget Sound (Discovery Bay to the west, Port Townsend and Oak Bays to the east), and as withdrawals from wells. A more detailed representation of the groundwater budget of the model area is provided by the equation:

$$GWin + Rppt + Rsw + Rsec = \\ GWout + Dsw + Det + Dppg + \Delta \qquad (2)$$

where

$Rppt$ is recharge from precipitation;

Rsw is recharge from streams and lakes;,

$Rsec$ is secondary recharge (return flow);

Dsw is groundwater discharge to streams, lakes, springs, seeps, and out of the terrestrial part of the model toward Puget Sound;

Det is groundwater discharge by evapotranspiration; and

$Dppg$ is groundwater withdrawal from wells.

All water-budget components (table 8) can be quantified on the basis of the calibration except discharge by evapotranspiration. Because a calculation of net recharge was used in the model, the water lost to the system through direct evapotranspiration of groundwater is taken into account simply as part of that process, and Det is not calculated in the model. Using the entire model (including submarine areas) with simulation of current conditions ("Current Conditions" simulation), assuming the change in the volume of water stored within the system (ΔS) is zero, the contributions of the other budget components can be quantified. Substituting the calibrated-model values into the equation yields the values in table 8. The values in the water budget for the numerical simulation are close to those estimated and reported in table 3 (Jones and others, 2011).

The calibrated steady-state groundwater model budget can be used to make general observations about the flow system. Total flow through the groundwater system was about 26,496 acre-ft/yr in the model area. Precipitation was the primary source of water recharging the groundwater system (15,440 acre-ft/yr, or about 58 percent). Recharge from streams and lakes (6,073 acre-ft/yr) and groundwater inflow (4,089 acre-ft/yr) were 23 and 15 percent, respectively, whereas secondary recharge from domestic and agricultural return flows (894 acre-ft/yr) was about 3 percent. Groundwater discharge to streams, lakes, seeps, springs, and Puget Sound were the largest of the outflows from the groundwater system at -22,456 acre-ft/yr, representing about 85 percent of total outflow, whereas groundwater outflow across the northern model boundary (-2,541 acre-ft/yr) and groundwater pumpage (-1,498 acre-ft/yr) was about 10 and 6 percent, respectively.

Table 8.   Water-budget components for the terrestrial part of the calibrated steady-state model, Chimacum Creek Basin and vicinity, Jefferson County, Washington, 2001–09.

[**Inflow:** $GWin$, groundwater inflow; $Rppt$, recharge from precipitation; Rsw, recharge from streams and lakes; $Rsec$, secondary recharge; ΔS, change in groundwater storage. **Outflow:** $GWout$, groundwater outflow; Dsw, groundwater discharge to streams, lakes, springs, seeps; Det, groundwater discharge by evapotranspiration; $Dppg$, groundwater withdrawal from wells. Summing the rates of inflows and outflows may not equal the totals shown due to rounding. **Abbreviation:** acre-ft/yr, acre-feet per year]

Inflow	Rate (acre-ft/yr)	Outflow	Rate (acre-ft/yr)
$GWin$	4,089	$GWout$	-2,541
$Rppt$	15,440	Dsw	-22,456
Rsw	6,073	Det	Not calculated in model
$Rsec$	894	$Dppg$	-1,498
ΔS	0		
Total inflow	**26,496**	**Total outflow**	**-26,496**

Table 9 describes the net flows in the terrestrial part of the model (excluding submarine areas) and each of the components of the flow system for Chimacum Creek subbasin (fig. 1) and the subbasins adjacent to it: Quimper subbasin (areas that drain to Discovery Bay on the west), and to Indian-Marrowstone subbasin (areas that drain to Oak and Port Townsend Bays on the east side). Using the post-processing application ZONEBUDGET, the inflows and outflows in the different subbasins of the study area (fig. 1) were calculated (table 9).

Because table 9 shows net flows (inflow minus outflow), the totals here are smaller than in table 8, and examine the subareas of the model in more detail. Recharge is not affected by net calculations in that it flows only one direction, and so agrees closely between the tables 8 and 9 ($Rppt$ of 15,400 acre-ft/yr in table 8 compared to total precipitation recharge, 15,453 acre-ft/yr, in table 9, differences being due to small differences in the areal calculations. Secondary recharge is the same (894 acre-ft/yr), and withdrawals from wells are in reasonable agreement ($Dppg$ = -1,498 acre-ft/yr in table 8 compared to -1,506 acre-ft/yr in table 9, again the difference is due to the spatial extent of the calculation).

However, the net groundwater flow from the north is separated in table 8 into inflows ($GWin$ = 4,089 acre-ft/yr) and outflows ($GWout$ = -2,541 acre-ft/yr) for a net of 1,549 acre-ft/yr, which corresponds closely to the total groundwater inflow/outflow of +1,518 acre-ft/yr in table 9. Surface-water interactions also are treated differently in tables 8 and 9; the net flow to and from surface water in table 8 is -16,383 acre-ft/yr (Rsw = 6,073 acre-ft/yr and Dsw = -22,456 acre-ft/yr) and total of net terrestrial (surface water, seeps, and springs; -6,331 acre-ft/yr) and submarine (-10,022 acre-ft/yr) groundwater discharges in table 9 for a total of -16,353 acre-ft/yr.

Table 9 differentiates the various surface-water systems (Chimacum Creek, lakes, and springs) and drainage basins, and the direction of the submarine discharges (to the west or east). It also shows the net groundwater flows between the different subbasins, with net flow from Quimper to Chimacum Creek basin, and from Chimacum Creek basin to Indian-Marrowstone basin.

Description and Analyses of Model Simulations

Following the calibration of the steady-state model, simulations were developed to: (1) estimate the effects of various future conditions; (2) run particle tracking simulations to find sources and sinks associated with the model area in general, and Chimacum Creek and the PUD #1 Sparling well in particular, and; (3) assess the effects of agricultural well pumping and depth on streamflow in Chimacum Creek.

Description of Simulations

Four simulations were prepared to compare Current Conditions with three possible future conditions: (1) Probable Future Use based on population forecasts, (2) Full Beneficial Use of the PUD #1 water rights, and (3) installation of a Sanitary Sewer system in the Urban Growth Area (UGA).

Current Conditions

This simulation uses 2009 pumping and return flows for all drinking water and agricultural uses, with recharge based on National Weather Service defined "normal" precipitation (average of precipitation for 1981–2010, National Oceanic and Atmospheric Administration, 2007). This simulation is used for comparison to the results of other simulations.

Probable Future Use

This is the Current Conditions simulation except that PUD #1 pumpage (and return flows) increases according to Golder Associates (2010) estimates, which in turn are based on estimated population growth, distributed proportionally to PUD #1 wells as envisioned by PUD #1. The total PUD #1 pumpage does not reach the 1,408 acre-ft/yr water-right limit in the Full Beneficial Use simulation.

Table 9.   Model-derived groundwater flow and comparable estimates, Chimacum Creek Basin and vicinity, Jefferson County, Washington.

[**Estimated flow:** From table 3. Total cell-to-cell flows in drainage subbasins within the terrestrial part of the model. Base case is steady state with "normal" precipitation (average of precipitation for 1981–2010) recharge and 2009 pumpage. Sign convention as used by MODFLOW: positive numbers indicate flow into aquifer; negative numbers indicate flow out of aquifer. Data in **bold** indicate comparable values. **Abbreviations:** acre-ft/yr, acre-feet per year; –, not applicable]

Component	Flow to subbasin (acre-ft/yr)			Total (acre-ft/yr)	Estimated flow (acre-ft/yr)
	Chimacum Creek	Quimper	Indian–Marrowstone		
Precipitation recharge	11,654	1,278	2,521	**15,453**	**15,600**
Return flows:					
Public supply	353	99	218	**669**	**592**
Self-supplied	84	26	38	**149**	**136**
Agricultural	76	0	0	**76**	**77**
Total	513	125	256	**894**	**805**
Total recharge	12,167	1,403	2,777	**16,347**	**16,400**
Wells:					
Public supply	-784	-72	-105	**-961**	**-833**
Self-supplied	-124	-38	-56	**-219**	**-200**
Agricultural (groundwater only)	-327	0	0	**-327**	**-329**
Total	-1,235	-110	-161	**-1,506**	**-1,362**
Surface-water systems:					
Chimacum Creek and tributaries	**-5,424**	0	0	-5,424	**-8,174**
Lakes	684	-16	-55	613	
Springs and seeps	-46	-722	-753	-1,521	
Total	-4,786	-737	-808	-6,331	
Groundwater inflow/outflow (along northern edge)	-603	3,114	-993	1,518	not estimated
Submarine discharges:					
Discovery Bay	0	-2,528	0	-2,528	not estimated
Port Townsend Bay/Oak Bay	-810	0	-6,684	-7,494	
Total	-810	-2,528	-6,684	-10,022	
Flow between subbasins:					
To/from Chimacum Creek	–	-1,141	5,874	4,733	not estimated
To/from Quimper	1,141	–	0	1,141	
To/from Indian/Marrowstone	-5,874	0	–	-5,874	
Total	-4,733	-1,141	5,874	0	
Flow discrepancy (inflow-outflow)	0	0	0	0	

Full Beneficial Use

This simulation is the Current Conditions simulation with full utilization of the PUD #1 water rights permits. Anticipated pumping rates for all PUD #1 wells were supplied by the PUD #1. Return flows in the "Quimper" service area increase at the same rate as the total pumpage from PUD #1 wells. Domestic and other public water withdrawals and return flows were increased according to population (and thus water consumption) estimates in Golder Associates (2010). Recharge was based on "normal" annual precipitation (based on National Weather Service average precipitation for 1981–2010, National Oceanic and Atmospheric Administration, 2007). No changes to agricultural withdrawals or return flows were made.

Sanitary Sewer

This simulation is the same as the Probable Future Use simulation, except that return flows are turned off within the UGA extent. This simulated the diversion of household waste water from septic systems to sanitary sewers, with the attendant loss of groundwater return flows from the septic systems.

Comparison of Current Conditions, Probable Future Use, Full Beneficial Use, and Sanitary Sewer Simulations

Probable Future Use, Full Beneficial Use, and Sanitary Sewer simulations represent future conditions. All these simulations increase domestic withdrawals and some PUD #1 withdrawals and return flows according to population (water consumption) estimates in Golder Associates (2010). Precipitation recharge remains constant, based on "normal" annual precipitation. The results from these simulations are compared in table 10. The maximum change in groundwater level, a 70.42 ft decrease, was between the Current Conditions and the Probable Future Use simulations in unit Bedrock (model layer 6), with a similar decrease in unit LA (model layer 4). Between the Probable Future Use and the Full Beneficial Use simulations, the maximum change was a 43.00 ft decrease in unit LA (model layer 4), from which the PUD #1 withdraws public water supply, with decreases also seen in adjacent units MC and LC (model layers 3 and 5). The Sanitary Sewer simulation showed only small maximum water level changes.

Table 10. Ranges of water-level changes beween simulations representing current and future conditions, Chimacum Creek Basin and vicinity, Jefferson County, Washington.

[**Hydrologic unit:** UC, Upper Confining unit; UA, Upper Aquifer unit; MC, Middle Confining unit; LA, Lower Aquifer unit; LC, Lower Confining unit; OE, Bedrock unit]

Hydrogeologic unit– model layer	Differences in water-level simulations (feet)		
	Probable Future Use (compared to current conditions)	Full Beneficial Use (compared to probable future use)	Sanitary Sewer (compared to probable future use)
UC–1	-0.65 to 0.81	-0.45 to 0.64	-0.16 to 0
UA–2	-1.00 to 0.10	-1.02 to 0.16	-0.20 to 0
MC–3	-16.42 to 2.16	-21.4 to 4.1	-0.75 to 0
LA–4	-62.74 to 1.79	-43.0 to 3.3	-0.22 to 0
LC–5	-18.01 to 0.89	-16.4 to 2.0	-0.65 to 0
OE–6	-70.42 to 2.68	-7.7 to 4.7	-0.37 to 0

Differences in selected water-budget components for the Probable Future Use simulation are shown in table 11. The Probable Future Use simulation involves an increase in pumpage for public water systems and self-supplied residences, based on growth in population, and an associated increase in return flows. The increase in well pumpage is 241 acre-ft/yr (shown as negative in table 11 because the flow is out from the aquifer) and the return flows increase by 130 acre-ft/yr (table 11). Consumptive use is the net effect of withdrawals and return flows, and this net effect is a change of -111 acre-ft/yr.

The effects of this increase in consumptive use in the Probable Future Use simulation are shown in table 11. Most of the increase (74 acre-ft/yr, or 67 percent) comes from reduced discharge to Chimacum Creek and its tributaries. Most of the remaining 37 acre-ft/yr of increased consumptive use comes from additional groundwater flow from the Indian-Marrowstone subbasin. Differences in simulated groundwater altitude between Current Conditions and future conditions (Probable Future Use) are shown in figure 13. The greatest changes in water levels are small decreases (around one-half foot) in units LA (model layer 4) and Bedrock (model layer 6). This is primarily near the PUD #1 wells, but there is also lowering in the area to the west where bedrock is shallow and not very productive.

The only difference between the Full Beneficial Use and Probable Future Use simulations is the greater pumping in PUD #1 wells and the corresponding return flows. Because the PUD #1 wells pump primarily from hydrogeologic unit LA (model layer 4), the greatest drawdown (table 10) between the two simulations is in that unit, as well as in the adjacent units MC and LC (model layers 3 and 5). Because of the greater return flows, there are areas where the water levels are higher in the Full Beneficial Use than in the Probable Future Use simulations (fig. 14). These rises are in the public water service area where the septic systems return this increased pumpage to the uppermost model layer.

The difference between the Sanitary Sewer simulation and the Probable Future Use simulation is the removal of return flows in the UGA in the Sanitary Sewer simulation. As a result there are only lower water levels in the Sanitary Sewers simulation (fig. 15) than in the Probable Future Use simulation, and nowhere are there increases in water level with the Sanitary Sewer simulation (table 10). The declines with installation of sanitary sewers are in the UGA, where return flows from septic systems currently elevate water levels.

Table 11. Comparison of selected water budget components for the Current Conditions and Probable Future Use simulations for Chimacum Creek model subbasin, Jefferson County, Washington.

[Flows equal inflow minus outflow; negative flows are out of the groundwater system or out of the subbasin. Column entries may not add exactly due to rounding. **Abbreviations:** acre-ft/yr, acre-feet per year; NA, not applicable]

Component	Simulation (acre-ft/yr)		
	Current Conditions	Probable Future Use	Change from Current Conditions
Recharge from precipitation	11,654	11,654	0
Return flows	513	643	130
Withdrawals from wells	-1,235	-1,476	-241
Consumptive use	-722	-833	-111
Chimacum Creek and tributaries	-5,420	-5,346	74
Lakes	684	684	1
Small streams and seeps	-46	-46	0
Groundwater inflow/outflow along northern study area boundary	-603	-605	-2
Puget Sound	-810	-809	2
Flow between subbasins	-4,737	-4,701	36
To Quimper	1,141	1,127	-15
To Indian–Marrowstone	-5,878	-5,828	51

A.

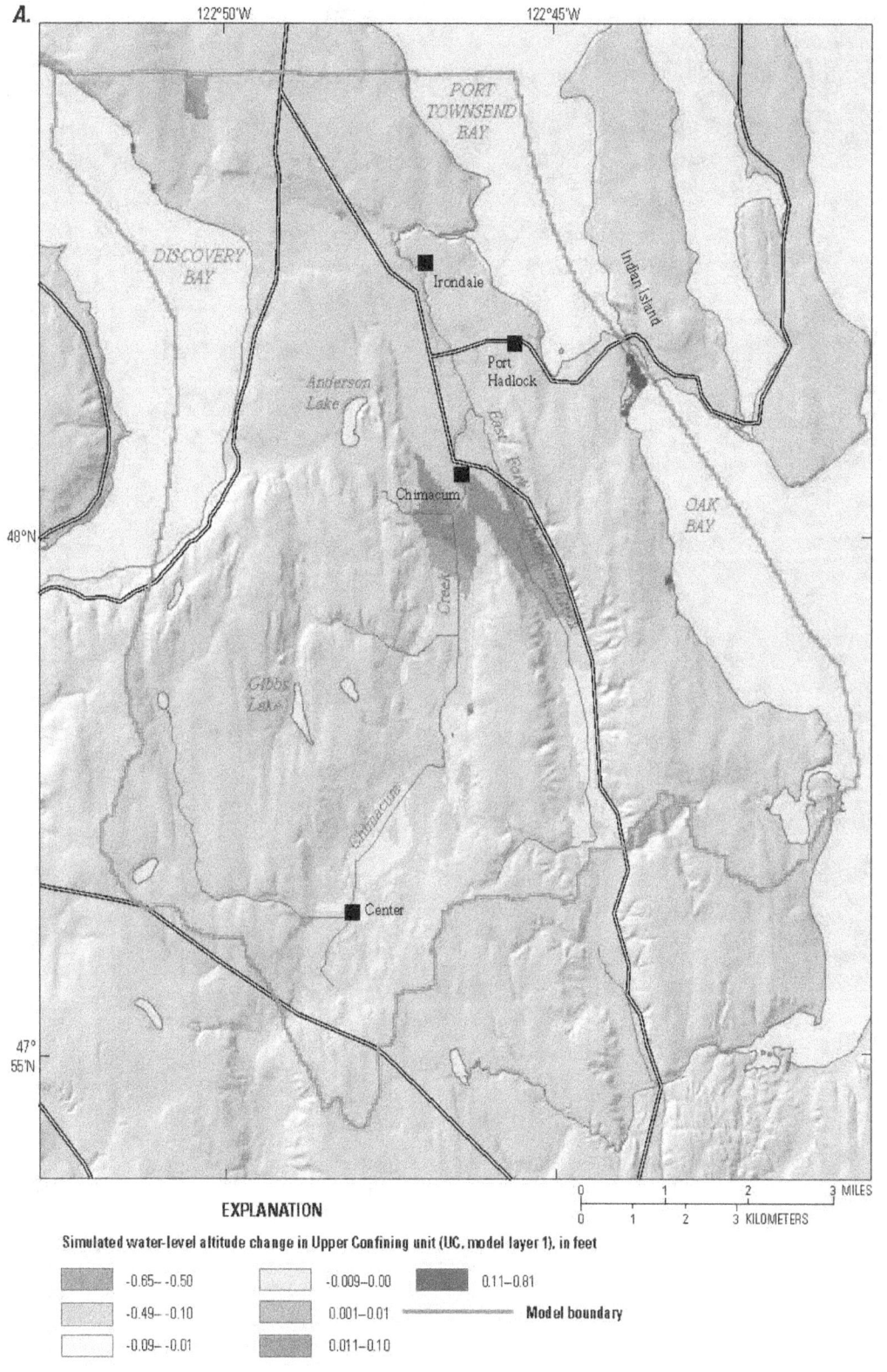

EXPLANATION

Simulated water-level altitude change in Upper Confining unit (UC, model layer 1), in feet

-0.65– -0.50	-0.009–0.00	0.11–0.81
-0.49– -0.10	0.001–0.01	Model boundary
-0.09– -0.01	0.011–0.10	

Figure 13. Simulated water-level altitude change in each hydrogeologic unit between the Current Conditions and Probable Future Use simulations, Chimacum Creek Basin and vicinity, Jefferson County, Washington.

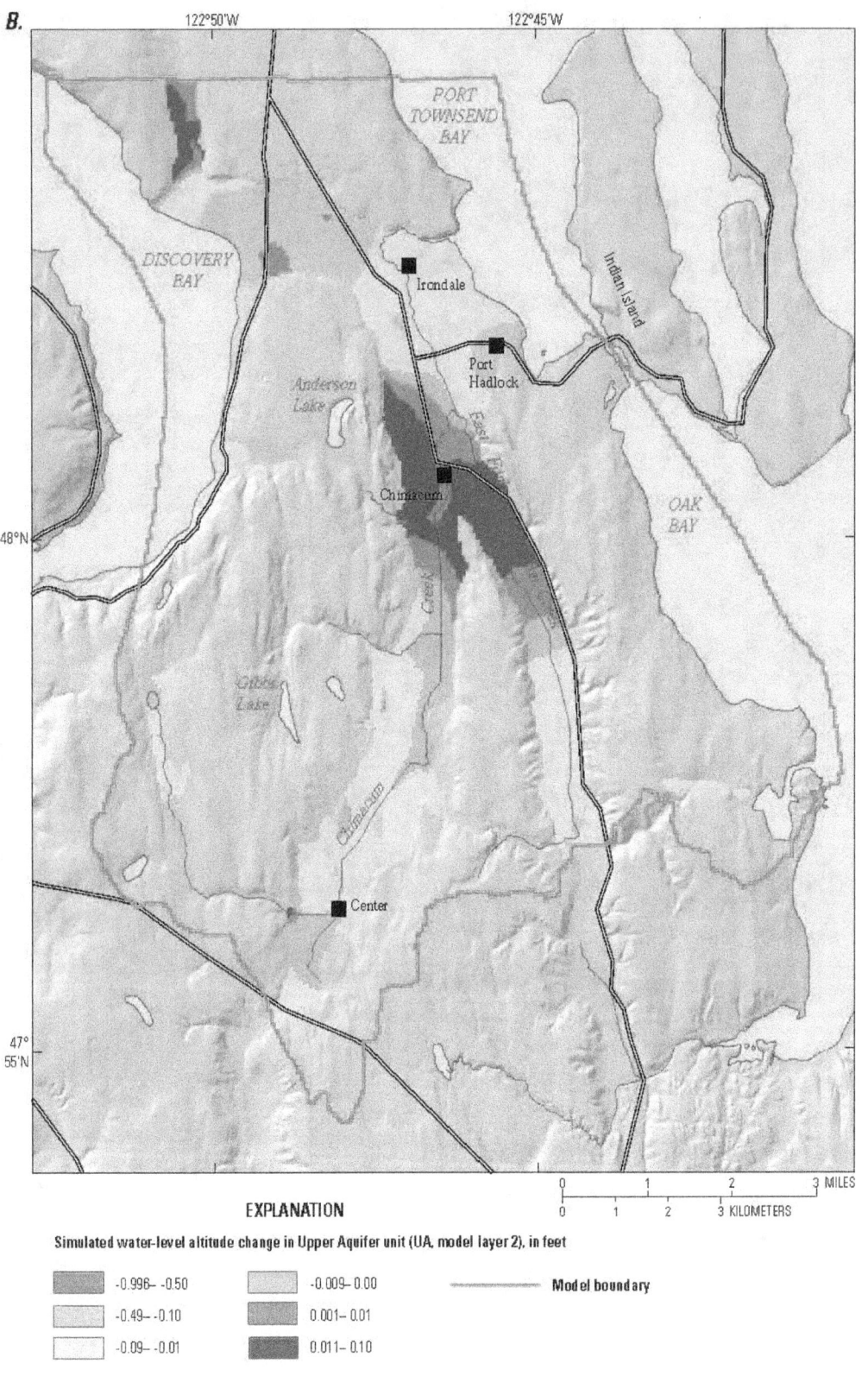

EXPLANATION

Simulated water-level altitude change in Upper Aquifer unit (UA, model layer 2), in feet

-0.996– -0.50	-0.009– 0.00	━━━ Model boundary
-0.49– -0.10	0.001– 0.01	
-0.09– -0.01	0.011– 0.10	

Figure 13.—Continued

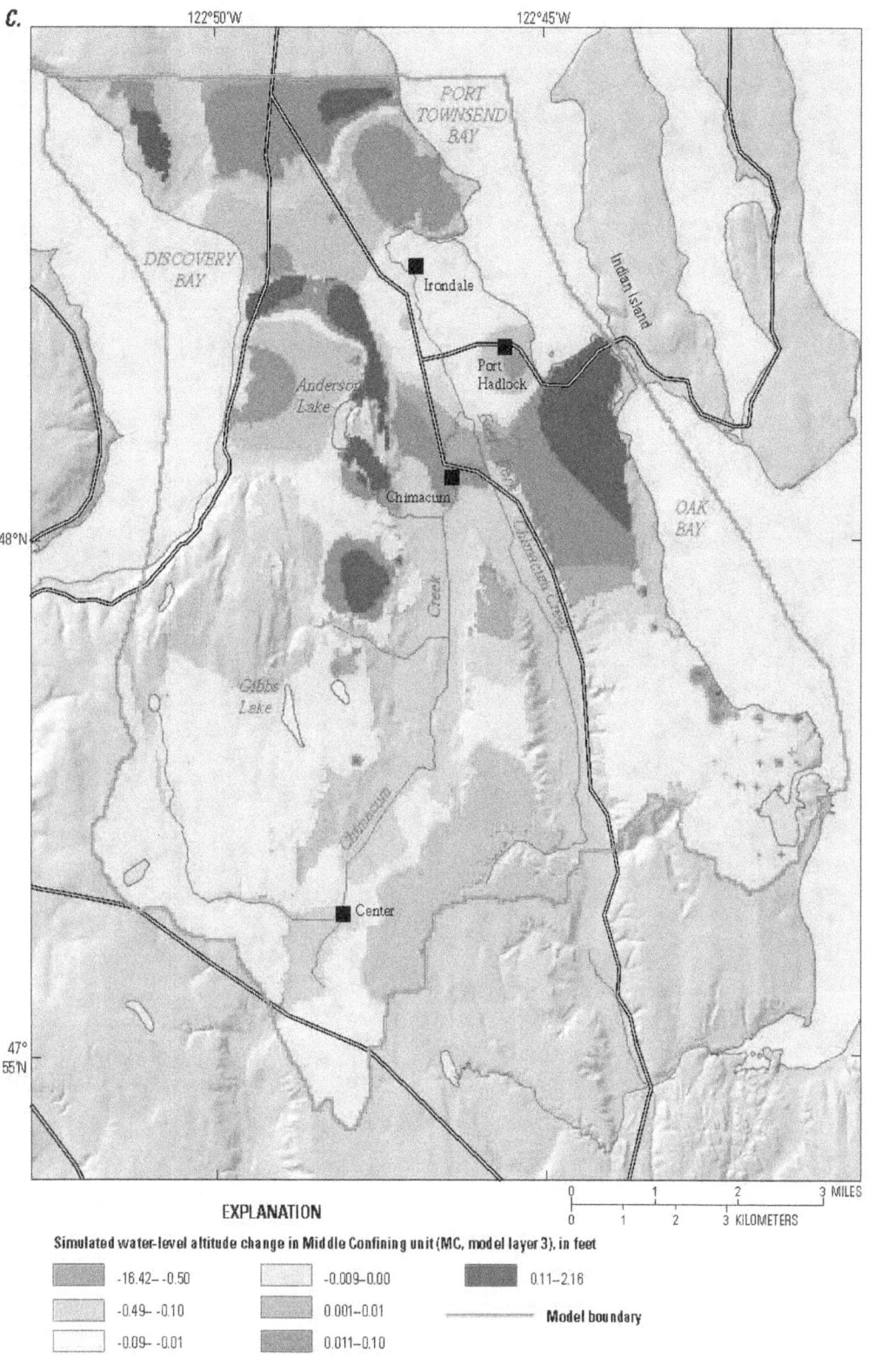

EXPLANATION

Simulated water-level altitude change in Middle Confining unit (MC, model layer 3), in feet

-16.42– -0.50

-0.49– -0.10

-0.09– -0.01

-0.009–0.00

0.001–0.01

0.011–0.10

0.11–2.16

Model boundary

Figure 13.—Continued

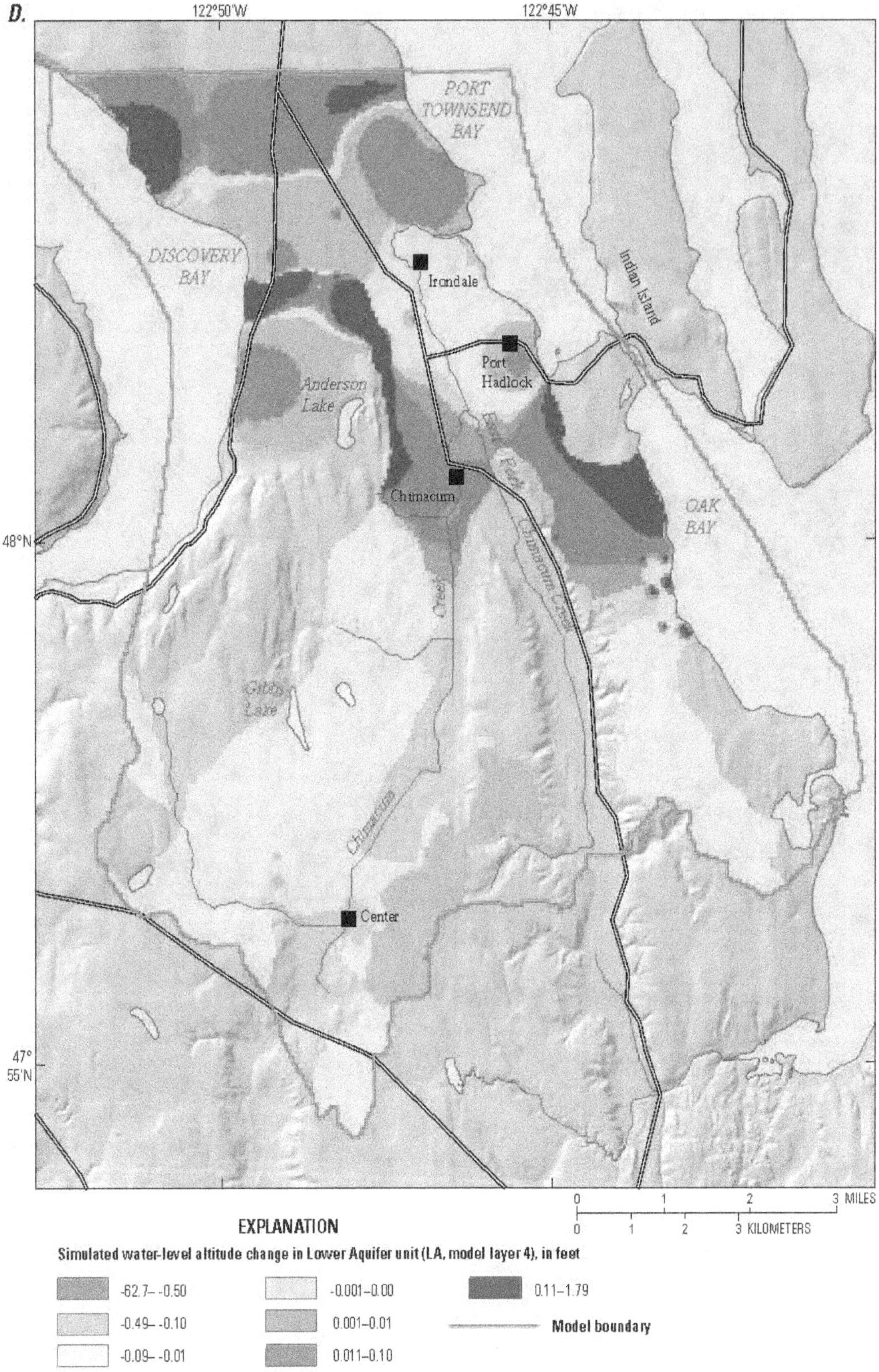
D.
122°50'W
122°45'W
PORT TOWNSEND BAY
DISCOVERY BAY
Irondale
Indian Island
Port Hadlock
Anderson Lake
Chimacum
OAK BAY
48°N
Gibbs Lake
Center
47° 55'N
0 1 2 3 MILES
0 1 2 3 KILOMETERS
EXPLANATION
Simulated water-level altitude change in Lower Aquifer unit (LA, model layer 4), in feet
-62.7– -0.50
-0.49– -0.10
-0.09– -0.01
-0.001–0.00
0.001–0.01
0.011–0.10
0.11–1.79
Model boundary

Figure 13.—Continued

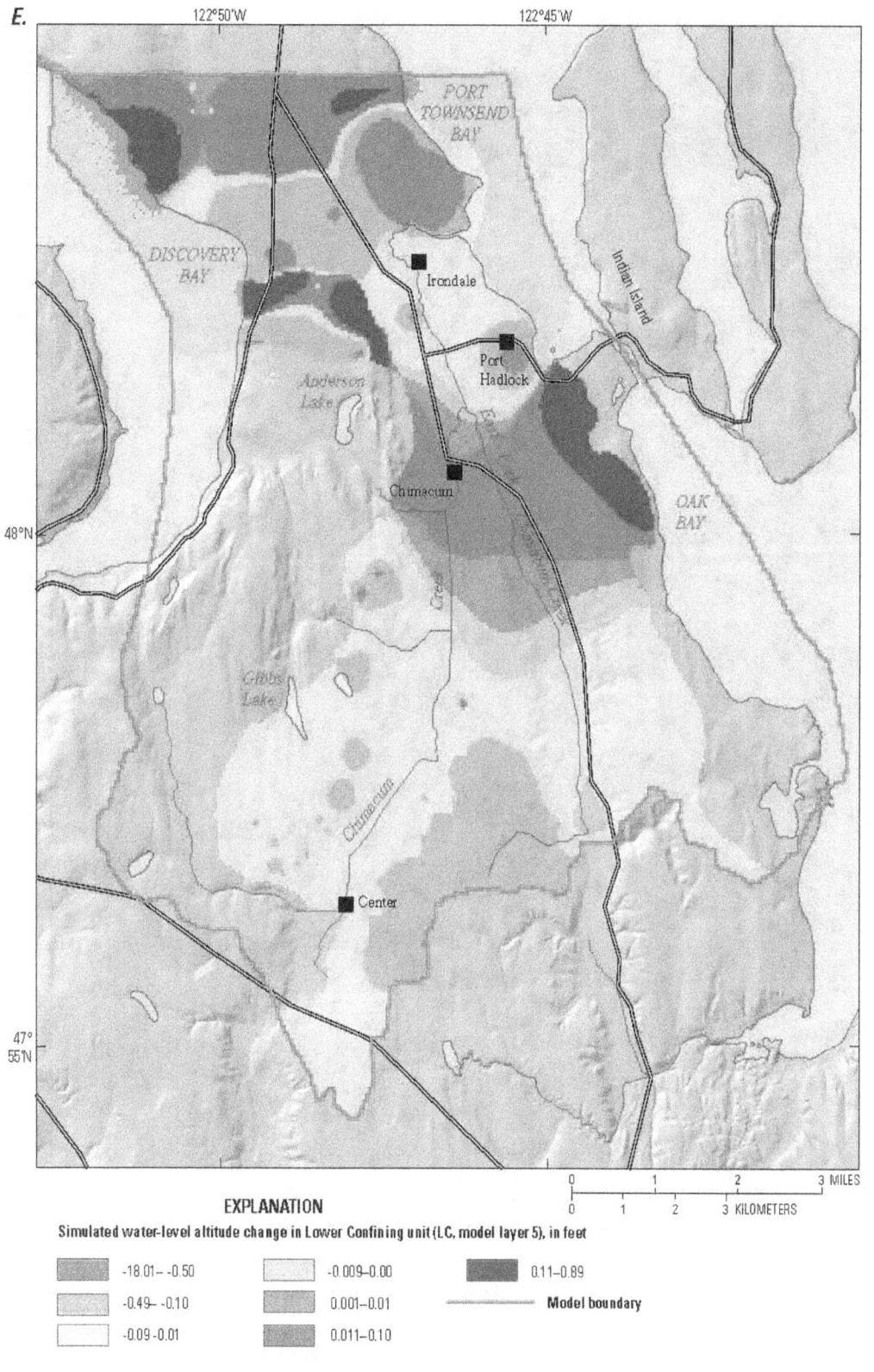

E.

EXPLANATION

Simulated water-level altitude change in Lower Confining unit (LC, model layer 5), in feet

- -18.01 – -0.50
- -0.49 – -0.10
- -0.09 - 0.01
- -0.009 – 0.00
- 0.001 – 0.01
- 0.011 – 0.10
- 0.11 – 0.89
- Model boundary

Figure 13.—Continued

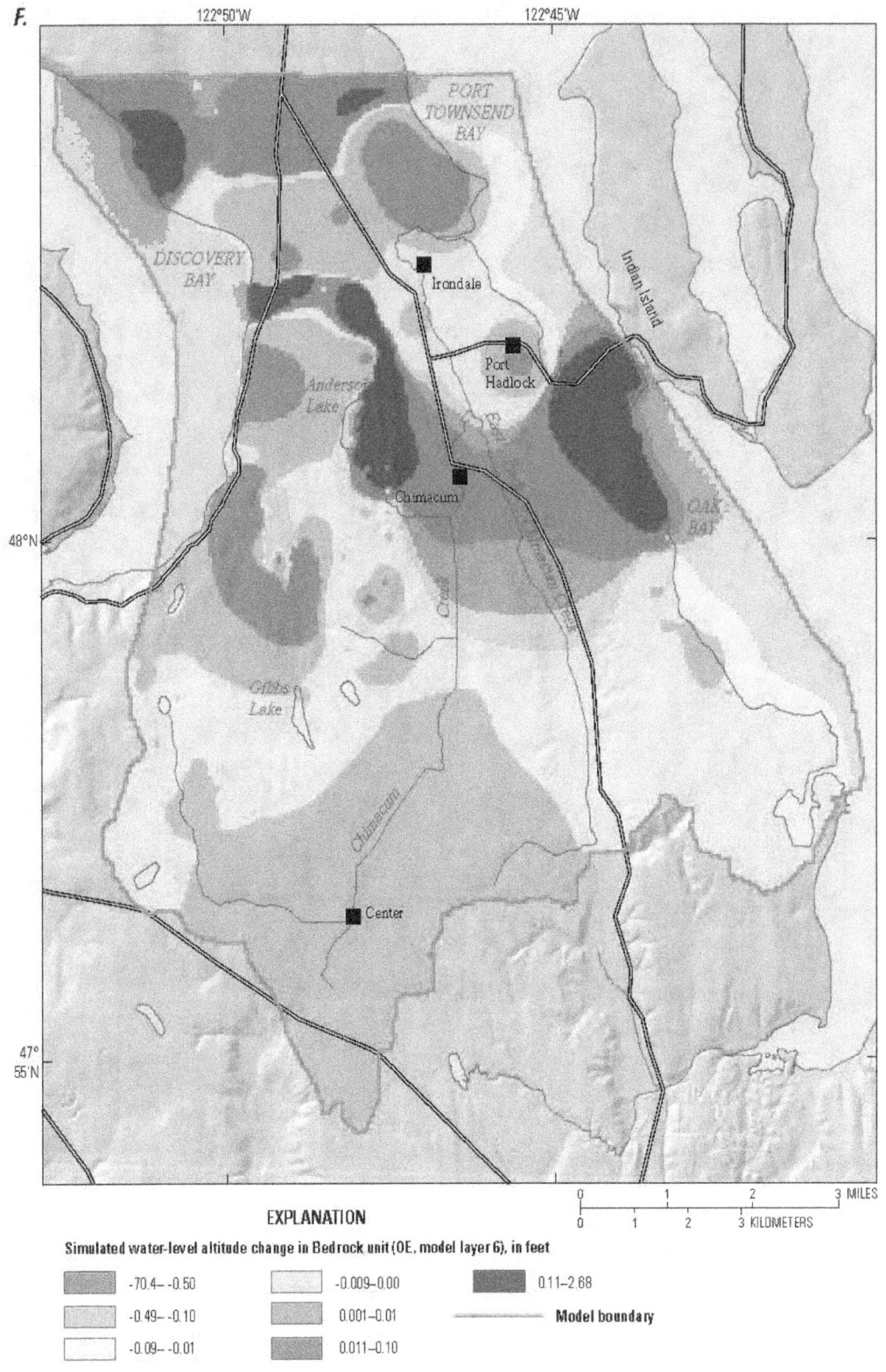

F.

EXPLANATION

Simulated water-level altitude change in Bedrock unit (OE, model layer 6), in feet

-70.4– -0.50	-0.009–0.00	0.11–2.68
-0.49– -0.10	0.001–0.01	Model boundary
-0.09– -0.01	0.011–0.10	

Figure 13.—Continued

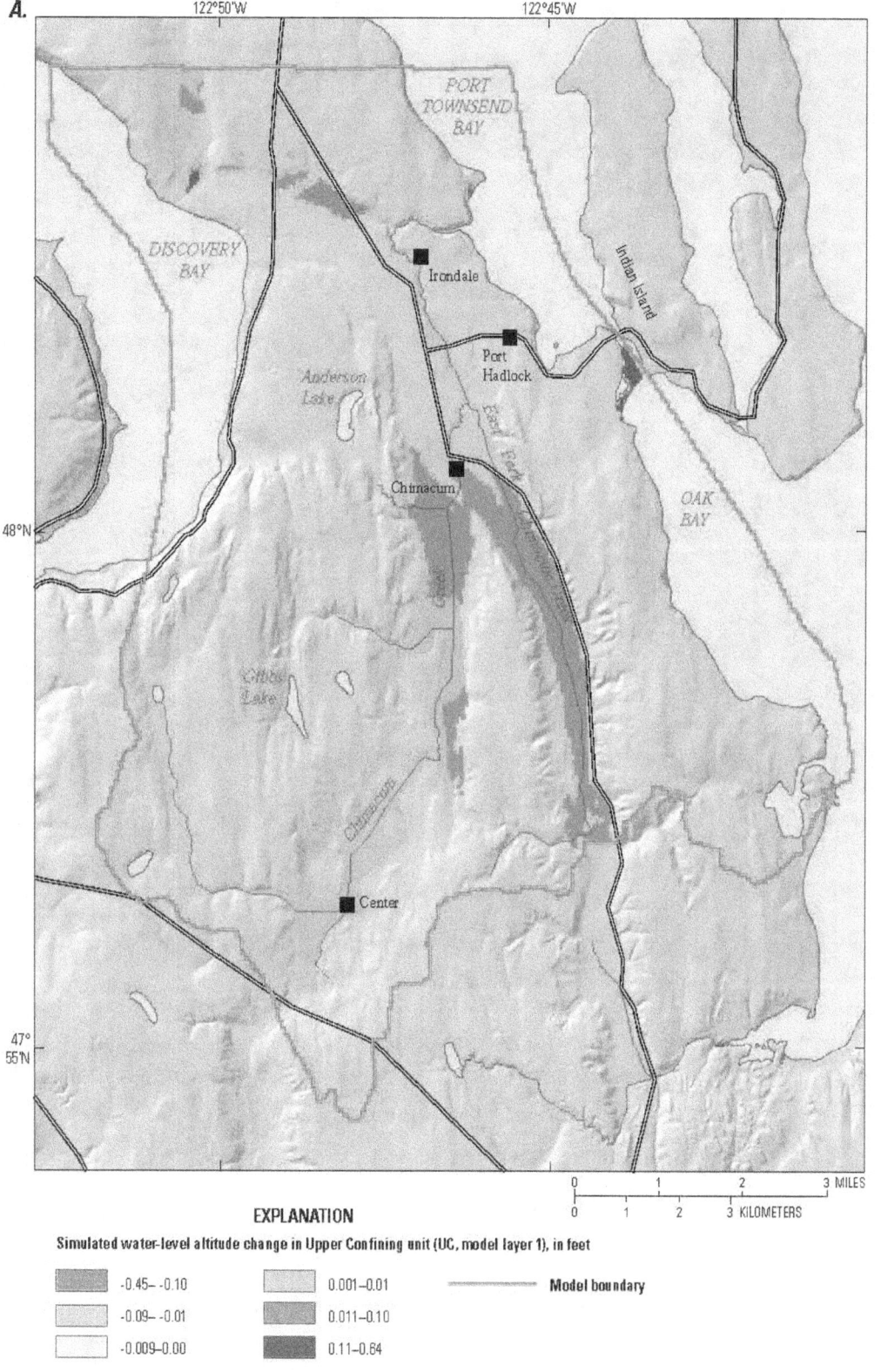

Figure 14. Simulated water-level altitude change in each hydrogeologic unit between Probable Future Use and Full Beneficial Use simulations, Chimacum Creek Basin and vicinity, Jefferson County, Washington.

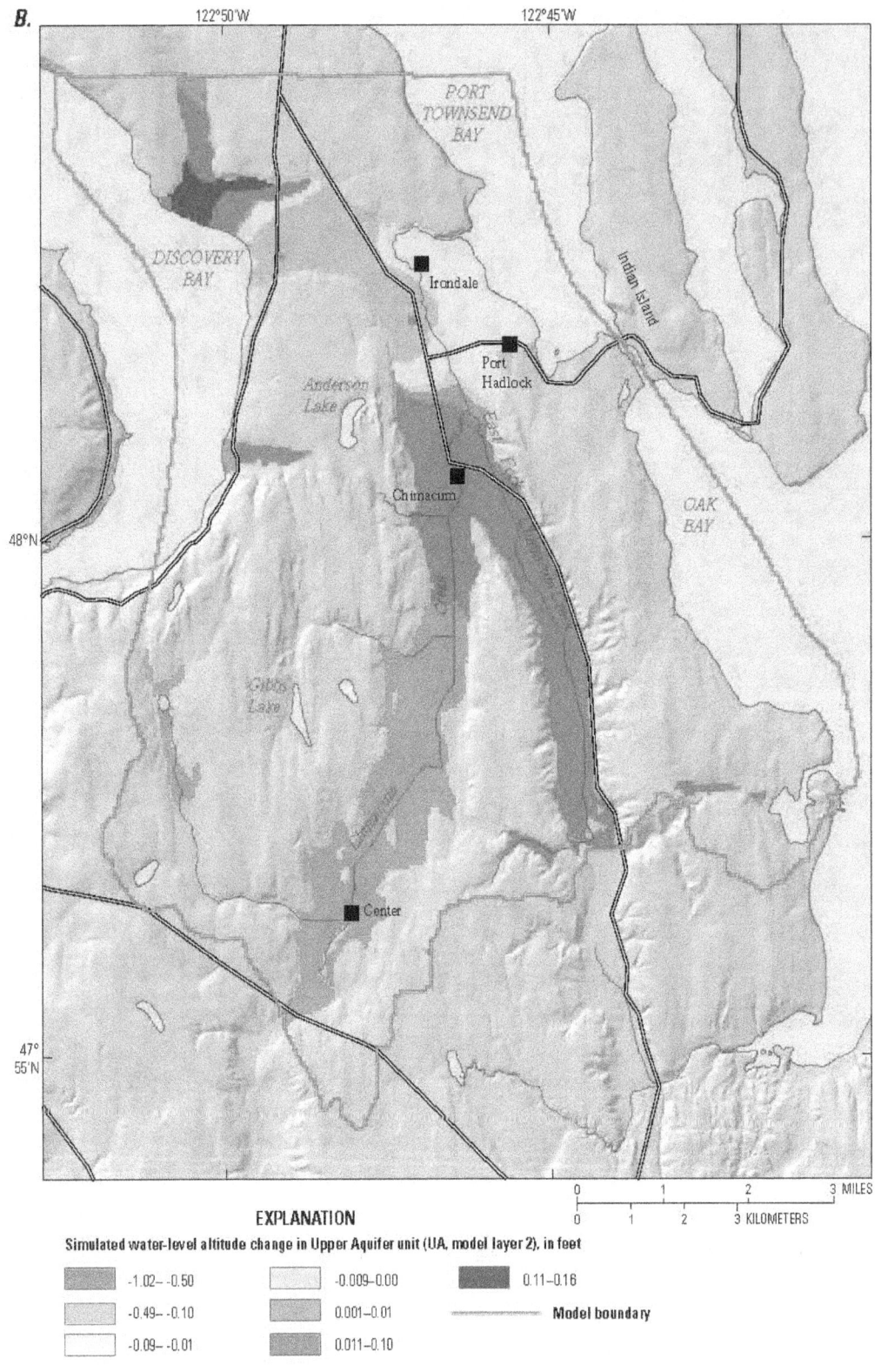

EXPLANATION

Simulated water-level altitude change in Upper Aquifer unit (UA, model layer 2), in feet

-1.02- -0.50	-0.009-0.00	0.11-0.16
-0.49- -0.10	0.001-0.01	Model boundary
-0.09- -0.01	0.011-0.10	

Figure 14.—Continued

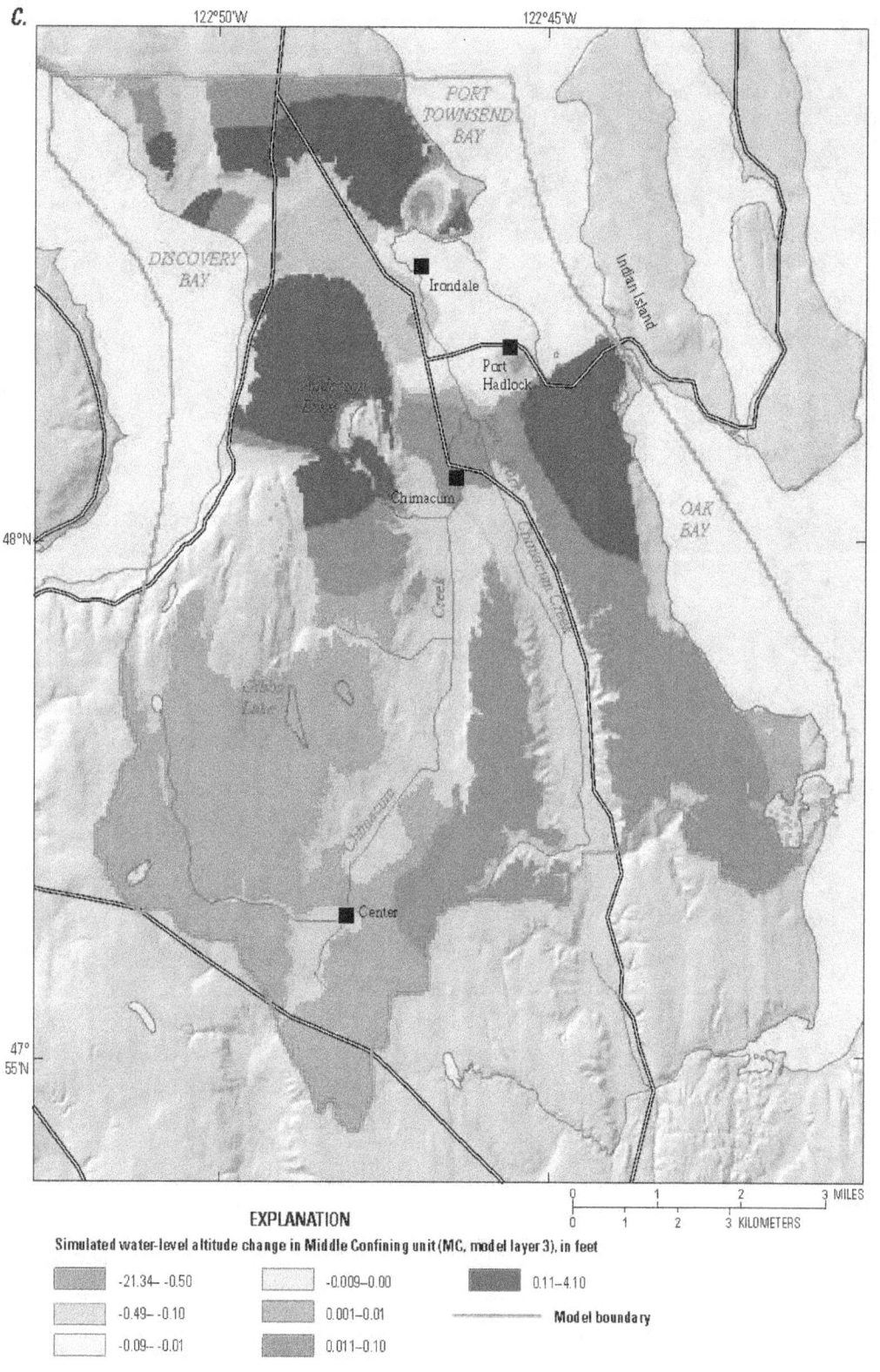

C.

EXPLANATION

Simulated water-level altitude change in Middle Confining unit (MC, model layer 3), in feet

	-21.34– -0.50		-0.009–0.00		0.11–4.10
	-0.49– -0.10		0.001–0.01		Model boundary
	-0.09– -0.01		0.011–0.10		

Figure 14.—Continued

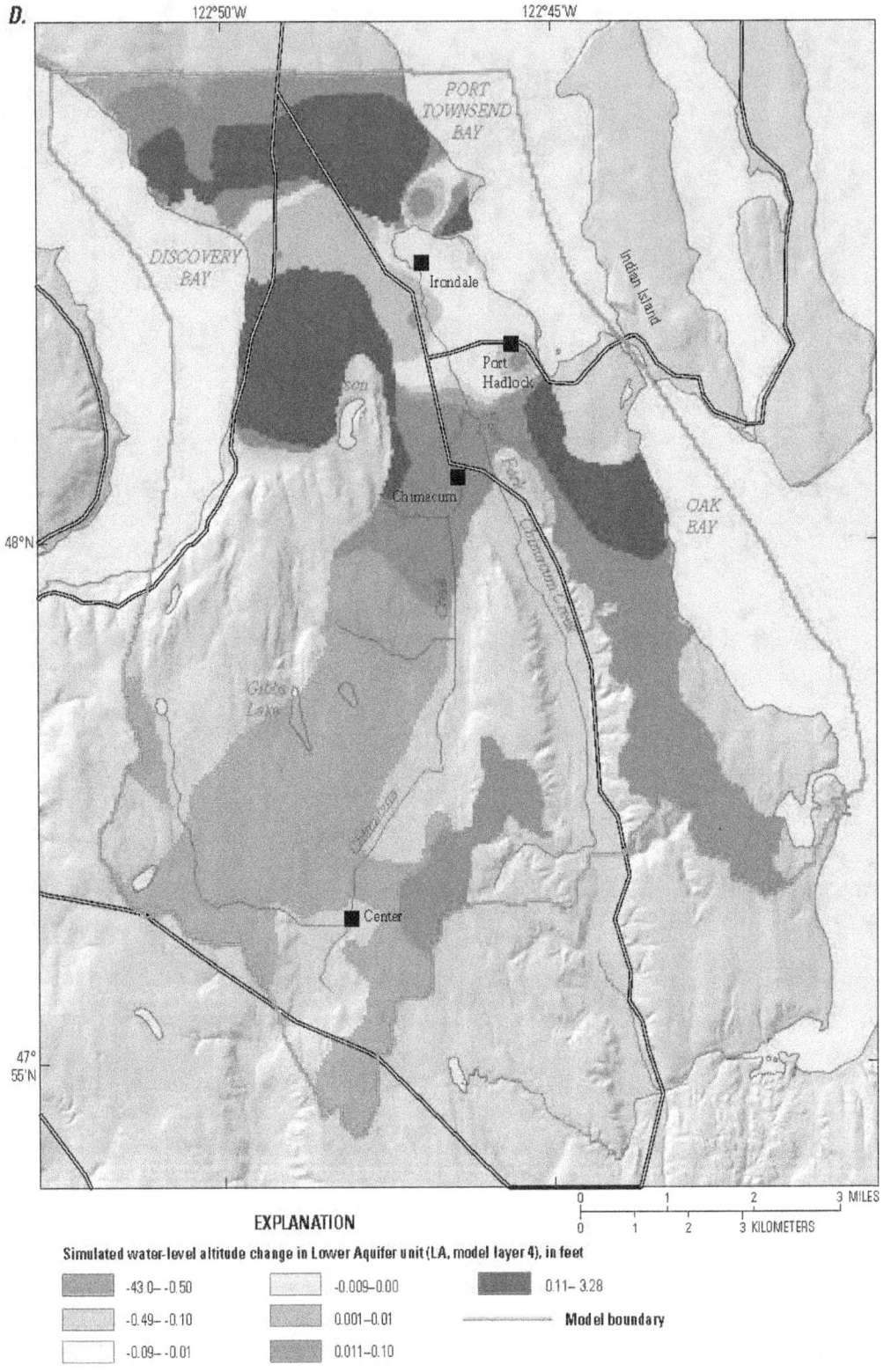

D.

EXPLANATION

Simulated water-level altitude change in Lower Aquifer unit (LA, model layer 4), in feet

-43.0 – -0.50	-0.009 – 0.00	0.11 – 3.28
-0.49 – -0.10	0.001 – 0.01	Model boundary
-0.09 – -0.01	0.011 – 0.10	

Figure 14.—Continued

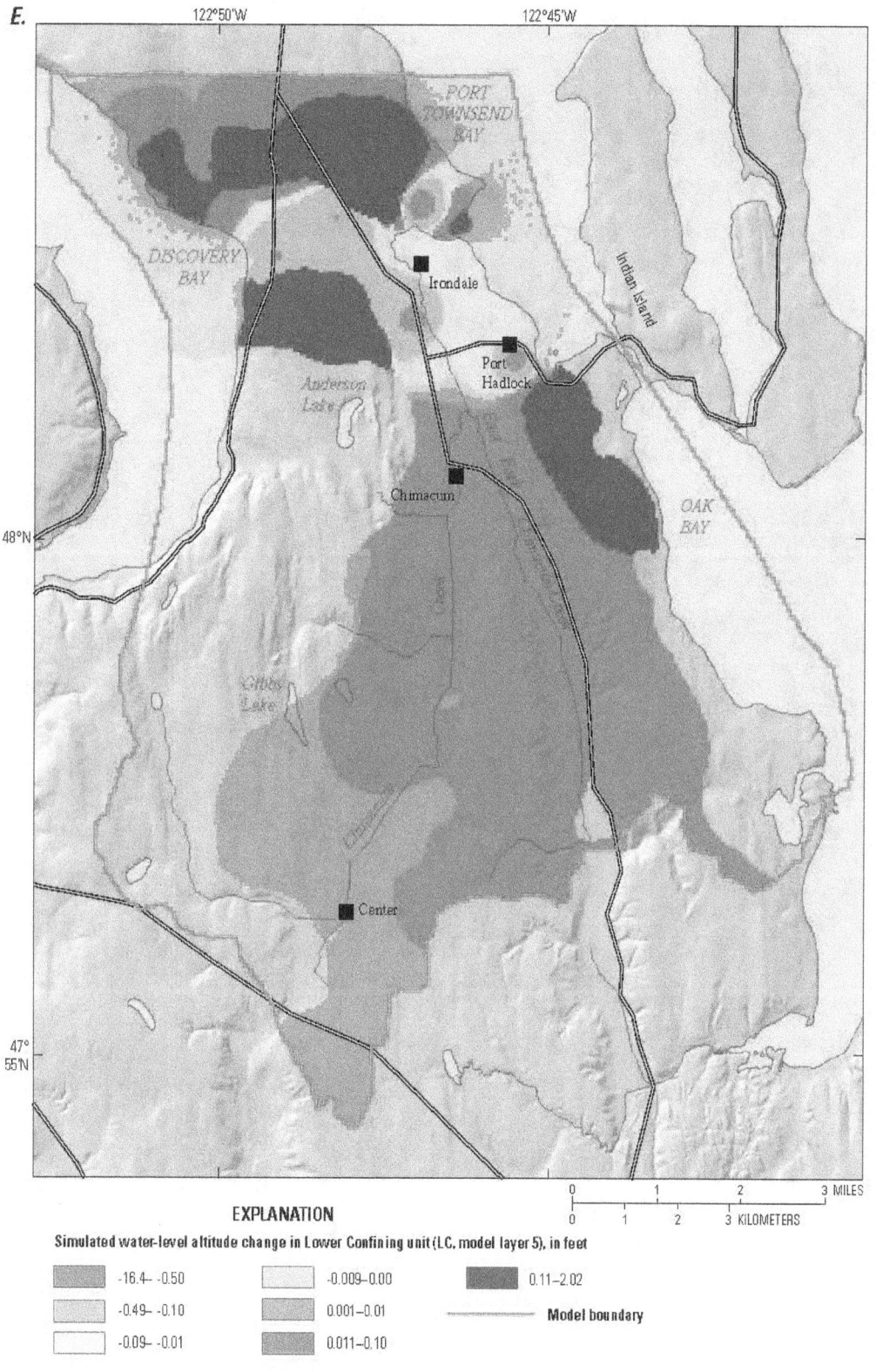

E.

EXPLANATION

Simulated water-level altitude change in Lower Confining unit (LC, model layer 5), in feet

	-16.4– -0.50
	-0.49– -0.10
	-0.09– -0.01

	-0.009–0.00
	0.001–0.01
	0.011–0.10

	0.11–2.02
	Model boundary

Figure 14.—Continued

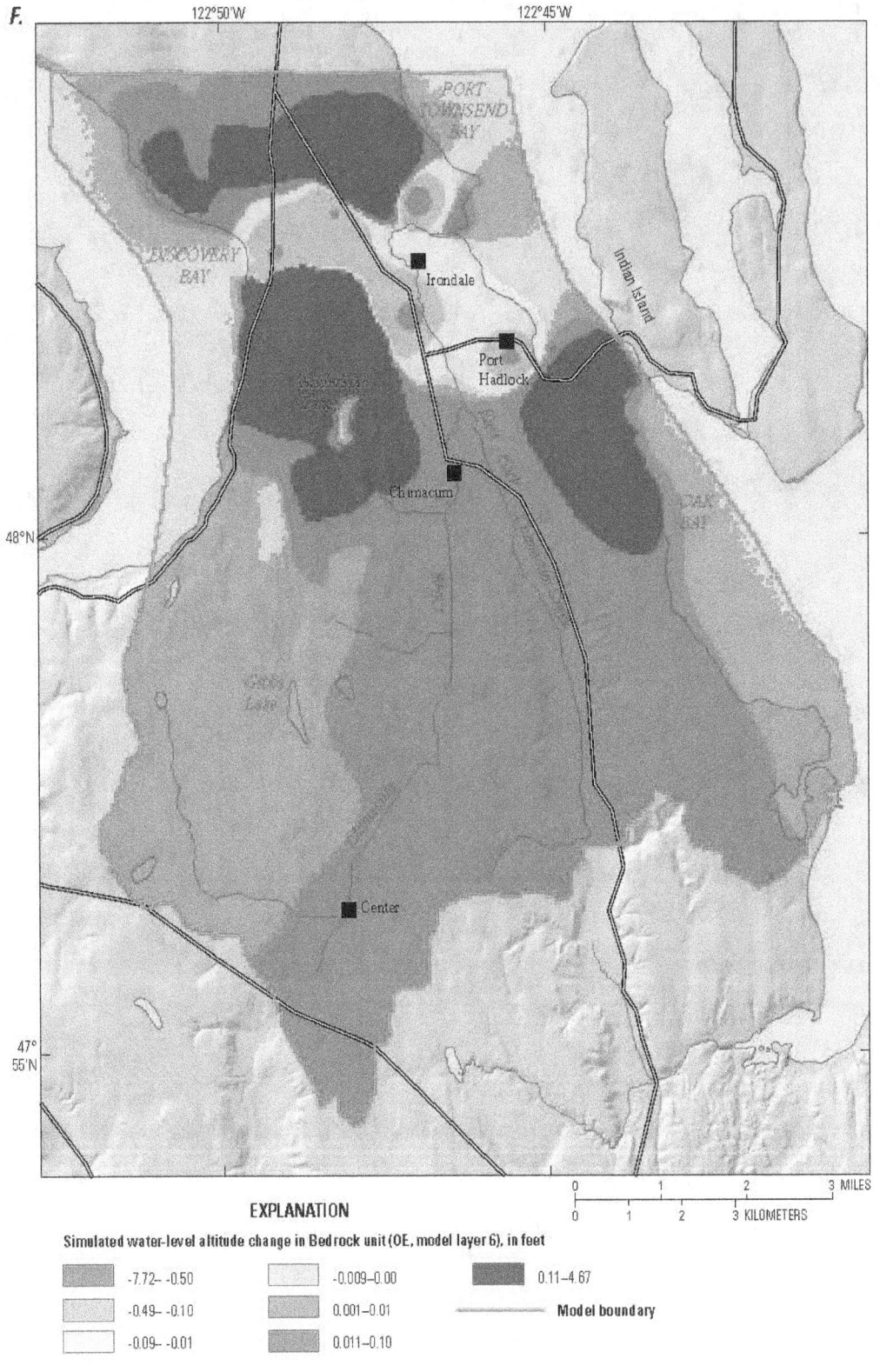

EXPLANATION

Simulated water-level altitude change in Bedrock unit (OE, model layer 6), in feet

■ -7.72– -0.50	■ -0.009–0.00	■ 0.11–4.67
■ -0.49– -0.10	■ 0.001–0.01	⸺ Model boundary
□ -0.09– -0.01	■ 0.011–0.10	

Figure 14.—Continued

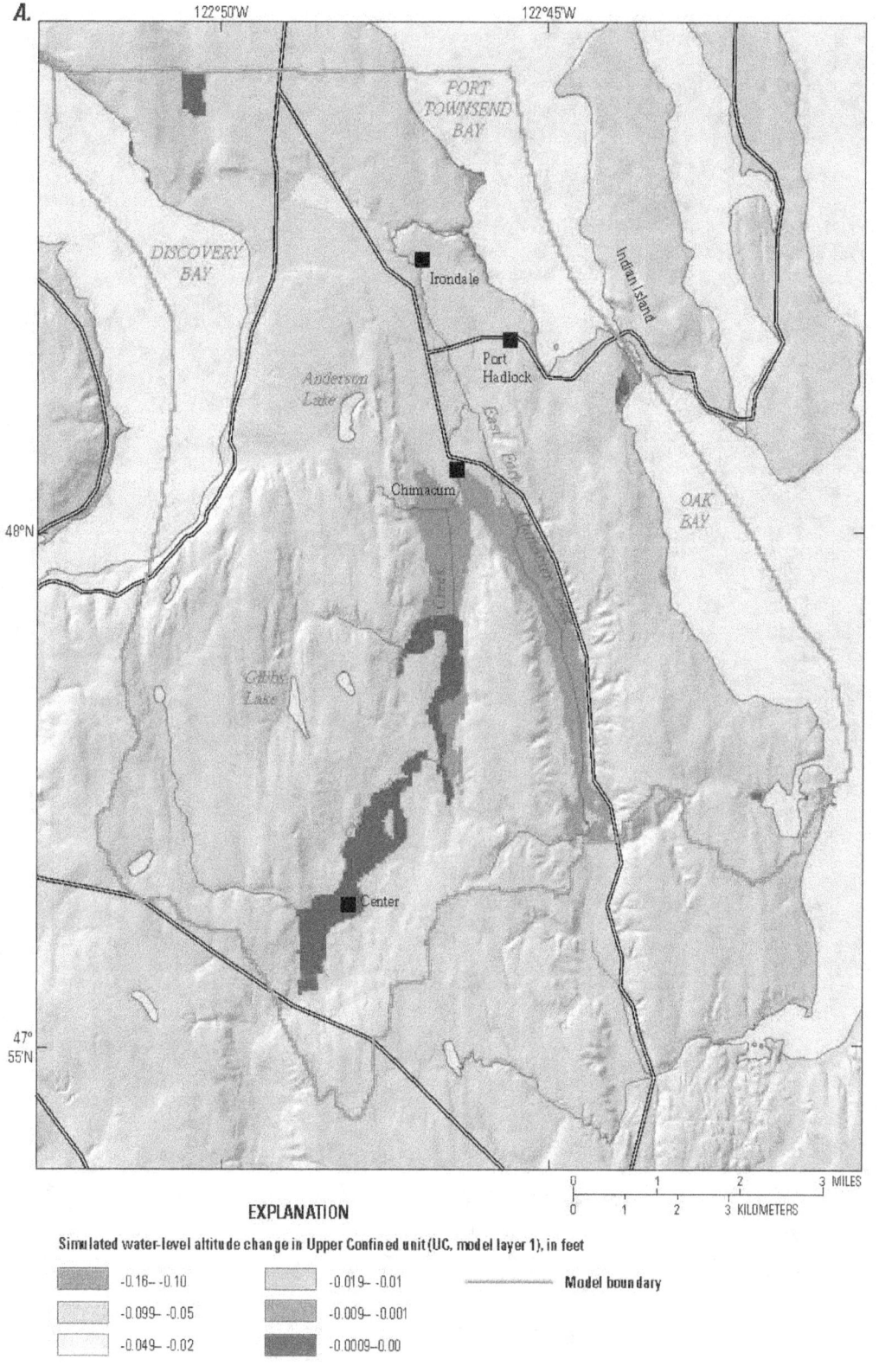

Figure 15. Simulated water-level altitude change in each hydrogeologic unit between Probable Future Use and Sanitary Sewer Simulations, Chimacum Creek Basin and vicinity, Jefferson County, Washington.

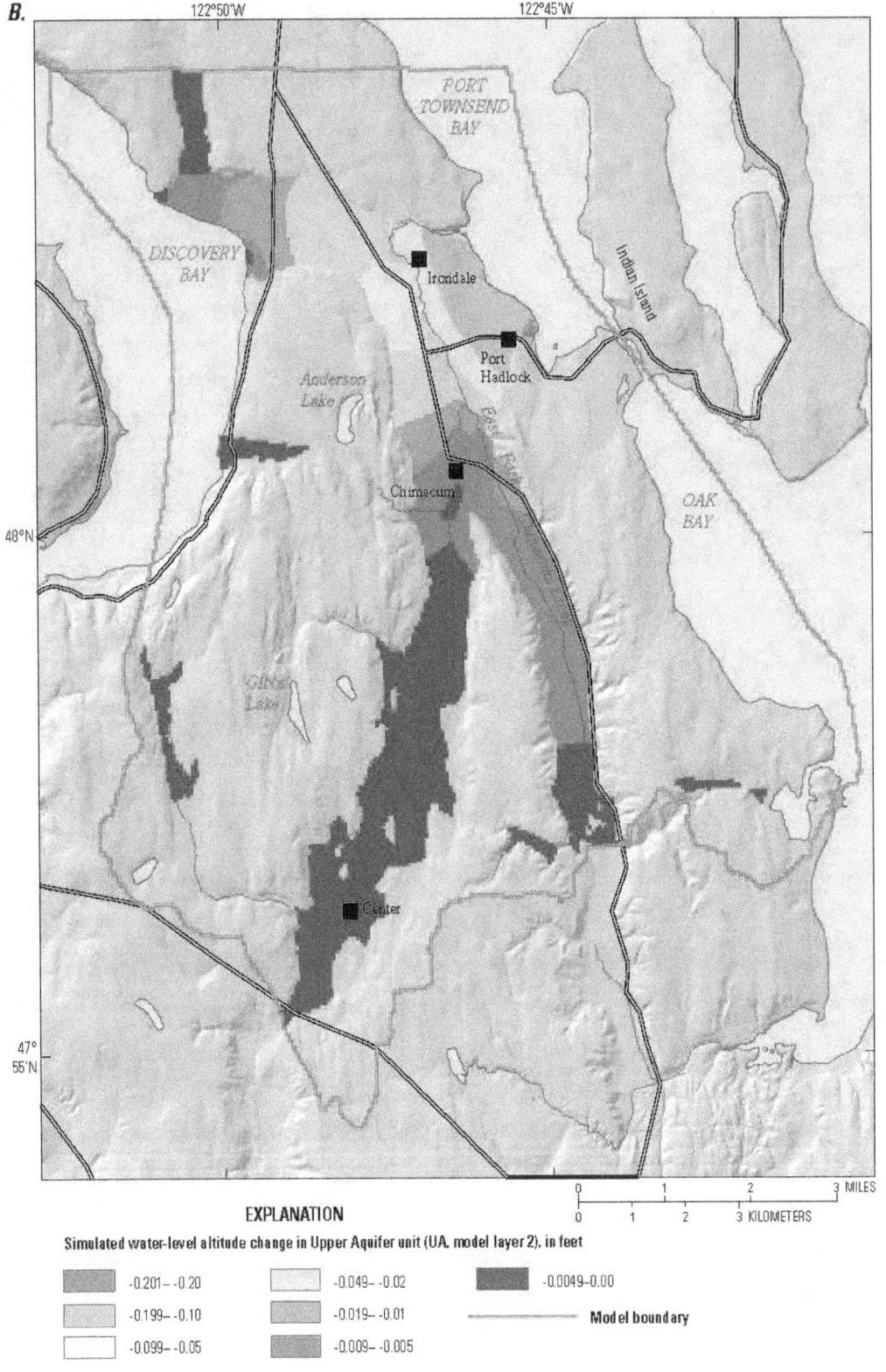

B.

EXPLANATION

Simulated water-level altitude change in Upper Aquifer unit (UA, model layer 2), in feet

-0.201– -0.20	-0.049– -0.02	-0.0049–0.00
-0.199– -0.10	-0.019– -0.01	Model boundary
-0.099– -0.05	-0.009– -0.005	

Figure 15.—Continued

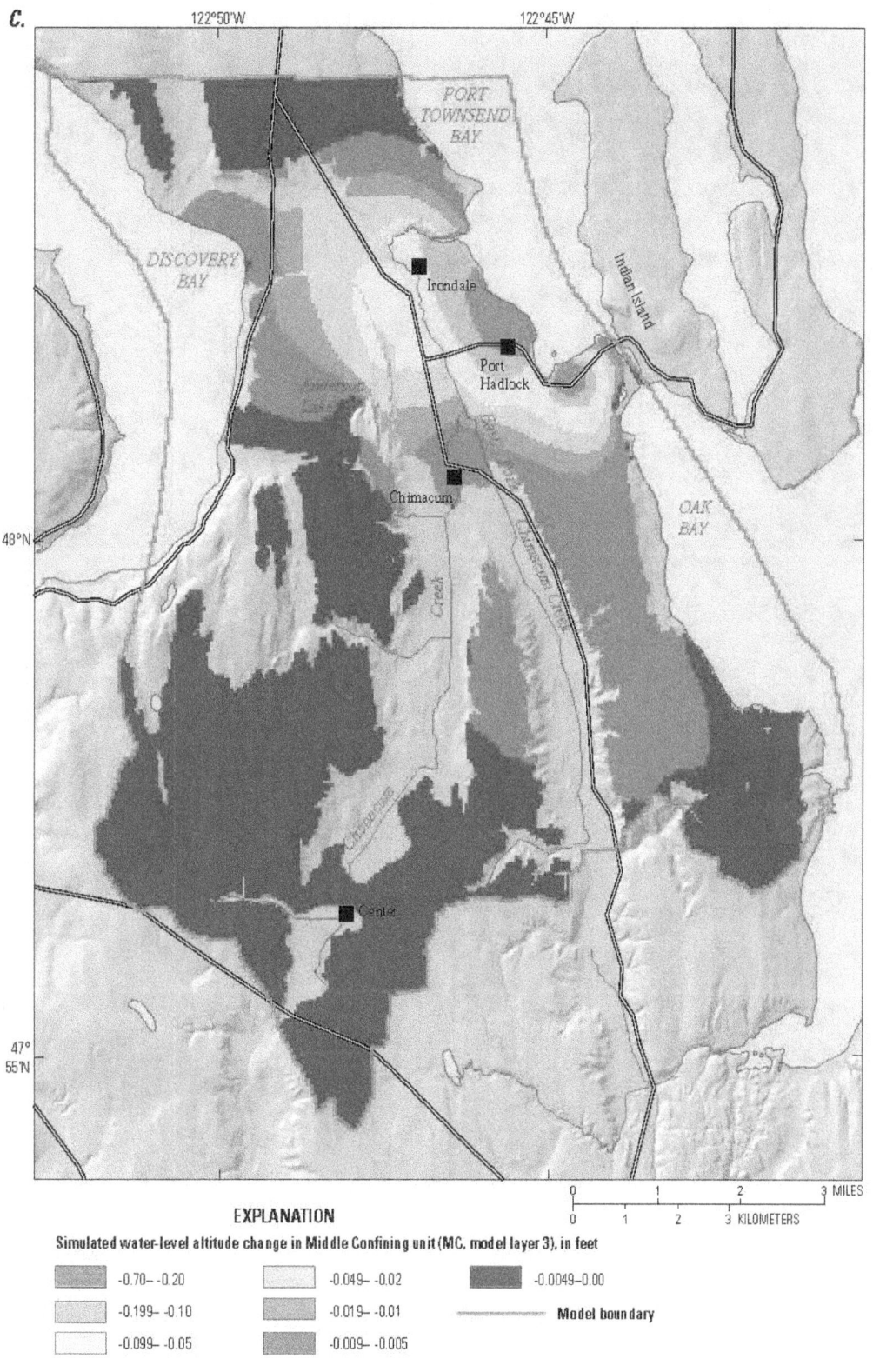

C.

EXPLANATION

Simulated water-level altitude change in Middle Confining unit (MC, model layer 3), in feet

-0.70– -0.20	-0.049– -0.02	-0.0049–0.00
-0.199– -0.10	-0.019– -0.01	Model boundary
-0.099– -0.05	-0.009– -0.005	

Figure 15.—Continued

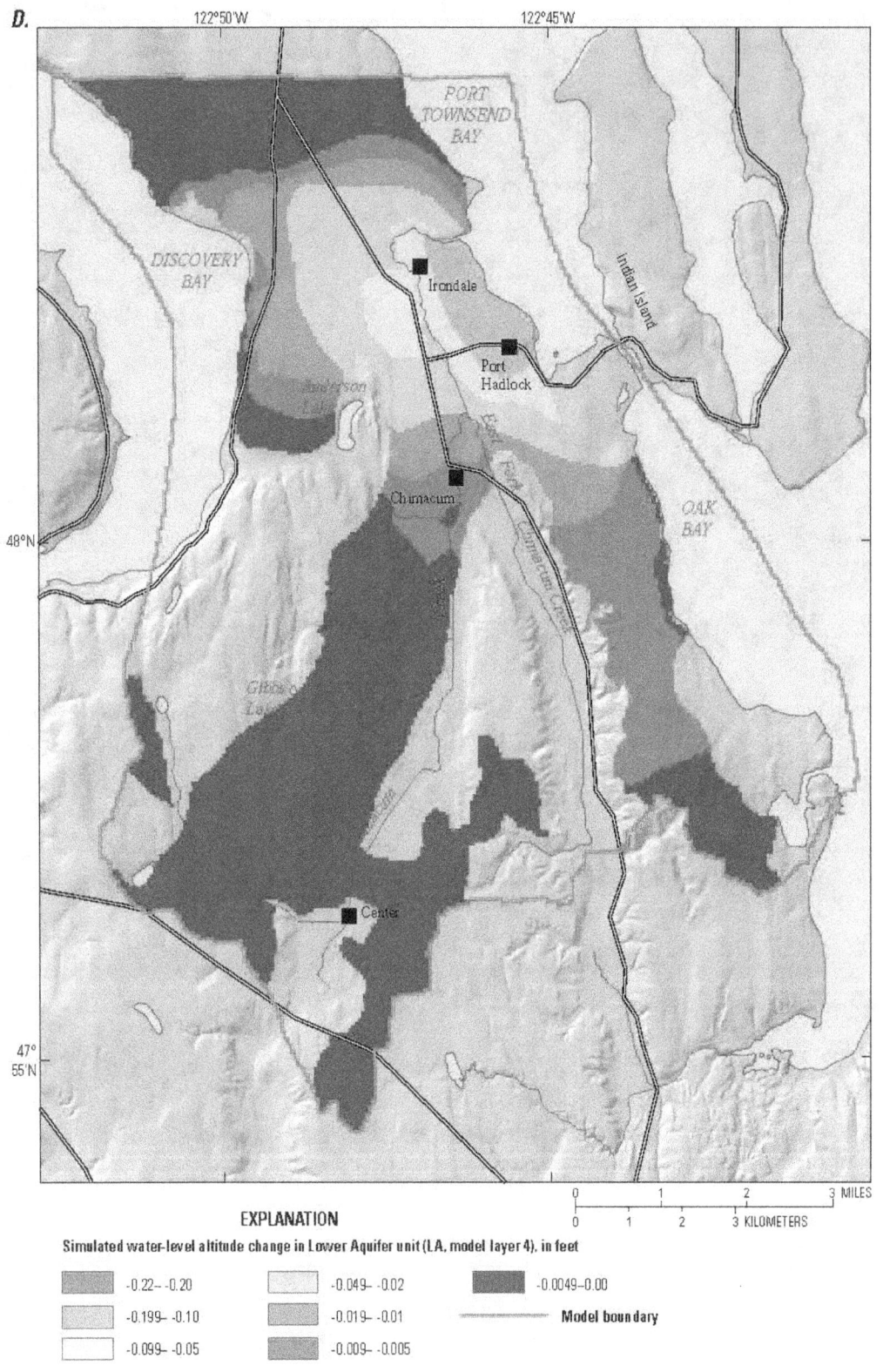

EXPLANATION

Simulated water-level altitude change in Lower Aquifer unit (LA, model layer 4), in feet

-0.22--0.20	-0.049--0.02	-0.0049--0.00
-0.199--0.10	-0.019--0.01	**Model boundary**
-0.099--0.05	-0.009--0.005	

Figure 15.—Continued

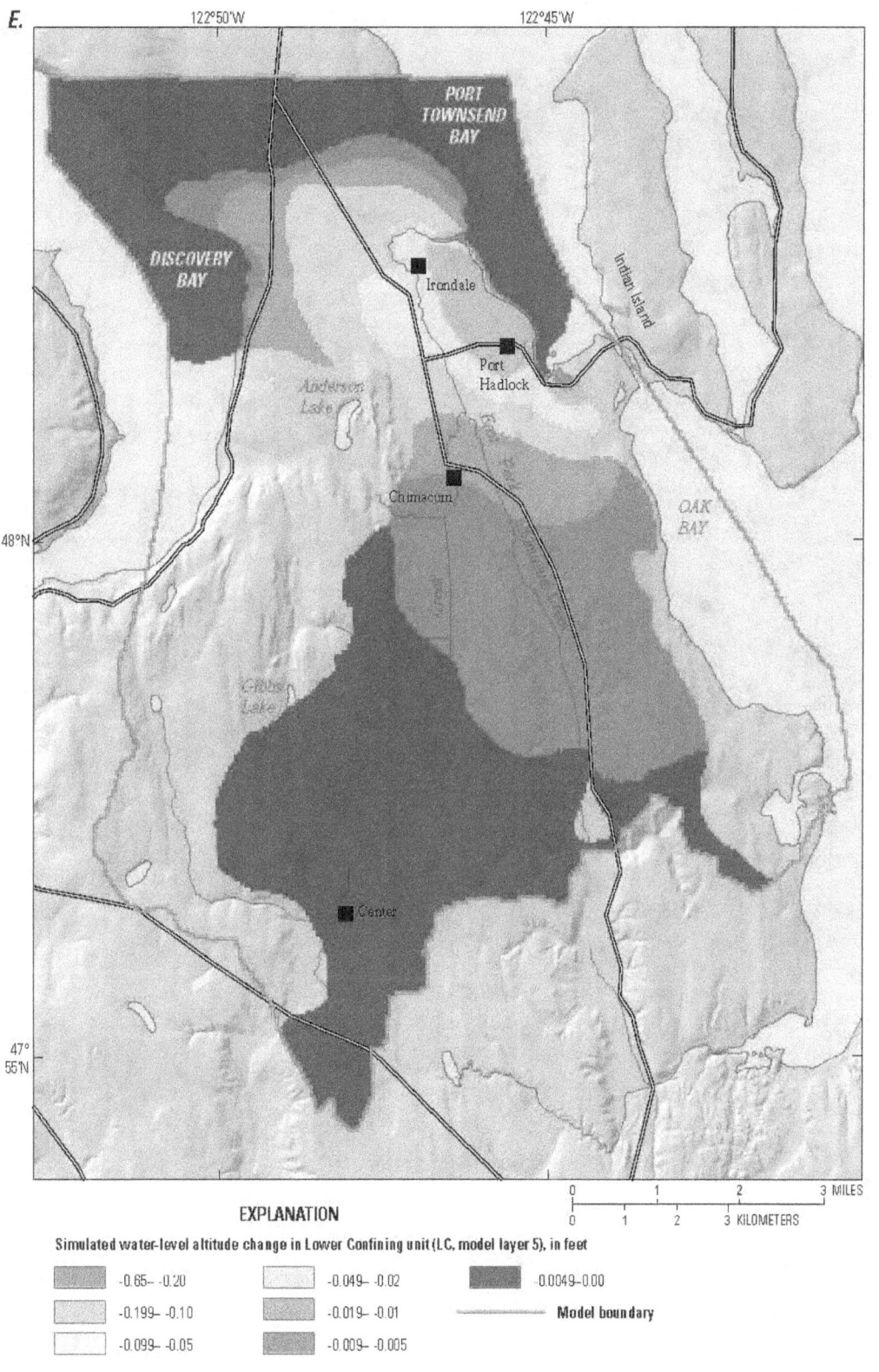

E.

EXPLANATION

Simulated water-level altitude change in Lower Confining unit (LC, model layer 5), in feet

-0.65– -0.20	-0.049– -0.02	-0.0049–0.00
-0.199– -0.10	-0.019– -0.01	Model boundary
-0.099– -0.05	-0.009– -0.005	

Figure 15.—Continued

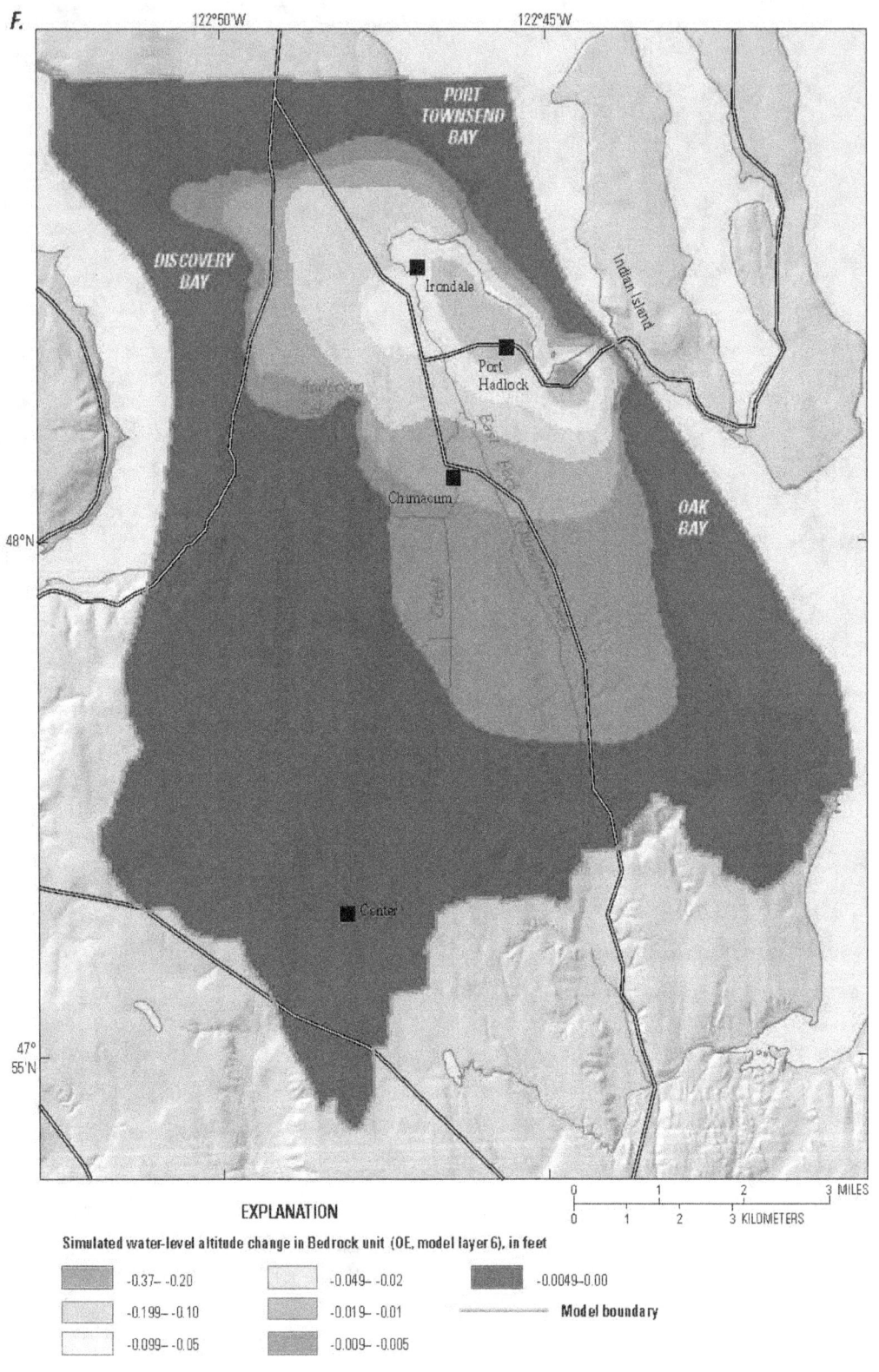

EXPLANATION

Simulated water-level altitude change in Bedrock unit (OE, model layer 6), in feet

-0.37– -0.20	-0.049– -0.02	-0.0049–0.00
-0.199– -0.10	-0.019– -0.01	Model boundary
-0.099– -0.05	-0.009– -0.005	

Figure 15.—Continued

The changes in selected water-budget components for the Full Beneficial Use and Sanitary Sewer simulations are shown in table 12. The Full Beneficial Use simulation increases consumptive use by 198 acre-ft/yr over the Probable Future Use simulation by increasing only PUD #1 pumpage and return flows (shown in the table as negative because there is more flow out of the aquifer). Almost all this increase (192 acre-ft/yr, or 97 percent) comes from reductions in seepage into Chimacum Creek streamflows, with a minor amount coming from flow toward Oak and Port Townsend Bays.

The effect of providing sanitary sewer service in the UGA, compared again to the Probable Future Use simulation is shown in table 12. Pumping in the Sanitary Sewer simulation is the same as in the Probable Future Use simulation, but the consumptive use increases because less return flow is provided to offset the pumping (the septic system return flows now go to the sanitary sewer). The additional increase in consumptive use is about 102 acre-ft/yr. All of the additional consumptive use is derived from reduced stream flow in Chimacum Creek. Additional streamflow loss is a result of increased flow toward Oak and Port Townsend Bays.

Table 12. Comparison of selected water budget components for the Probable Future Use, Full Beneficial Use, and Sanitary Sewer simulations, Chimacum Creek model subbasin, Jefferson County, Washington.

[Column entries may not add exactly due to rounding. **Abbreviations:** acre-ft/yr, acre-foot per year; NA, not applicable]

Component	Simulation (acre-ft/yr)				
	Probable Future Use	Full Beneficial Use	Change from Probable Future Use	Sanitary Sewer	Change from Probable Future Use
Recharge from precipitation	11,654	11,654	0	11,654	0
Return flows	643	810	167	541	-102
Withdrawals from wells	-1,476	-1,841	-365	-1,476	0
Consumptive use	-833	-1,031	-198	-935	-102
Chimacum Creek and tributaries	-5,346	-5,154	192	-5,235	111
Lakes	684	683	-1	684	0
Small streams and seeps	-46	-46	0	-46	0
Groundwater inflow/outflow along northern study area boundary	-605	-608	-3	-605	0
Puget Sound	-809	-808	1	-806	3
Flow between subbasins	-4,701	-4,690	11	-4,712	-11
To Quimper	1,127	1,107	-20	1,128	1
To Indian-Marrowstone	-5,828	-5,798	30	-5,840	-12

Flow Directions from Sources and to Sinks

The Current Conditions simulation was used to assess flow directions to and from model boundary conditions (streams, springs, and submarine discharge) using particle tracking with MODPATH (Pollock, 1994). MODPATH simulates the advective component of particle movement disregarding any contributions of dispersion and was used in both "forward" and "reverse" modes. In forward mode, particle starting locations are specified and MODPATH calculates where the flow field would take it (tracking down-gradient); in reverse mode, particle ending locations are specified and MODPATH calculates where it came from (tracking up-gradient). MODPATH was used for

- Forward tracking of particles scattered at select locations throughout the top model layer; this presents an overall picture of where particles from all over the model domain ultimately discharge.

- Source of water to the Sparling 2 well; these are reverse tracking MODPATH runs with particles originating at the Sparling 2 well in the Lower Aquifer unit (LA, model layer 4).

- Forward and reverse tracking of particles located in stream cells. Forward tracking shows where water leaking out of the creek to the aquifer discharges after flowing through the aquifer system; reverse tracking from stream cells shows where water flowing into the creek from the aquifer originates.

Forward particle tracking from selected locations throughout the model area is the most generally informative, whereas reverse tracking from stream cells allows more detailed evaluation of topics focusing on the stream (for example, where enhanced recharge increases stream discharge, or conversely, where pumping contributes to stream depletion), or on a well (size and location of a well capture zone). MODPATH results for forward tracking of particles distributed over selected locations throughout the uppermost model layer are shown in figure 16. The colors of the particle traces indicate the type of boundary condition at which the particles ultimately discharged.

Some of the particles that discharge at wells are discharging to domestic and agricultural wells, but not all flow paths to such wells are depicted because particles were not distributed densely enough to discharge at all wells regardless of withdrawal rate.

Driven by the higher precipitation and consequently higher recharge in the western and southwestern parts of the model, water table altitudes are the highest in those areas,

and flow directions indicated by the flow tracks generally move away from this area to the north and east. Particularly noteworthy is that groundwater in the southwest area almost exclusively flows to Chimacum Creek, indicating that this is the primary source of water to the stream. Particles originating in the northern part of the high groundwater table area, in the western part of the model, flow primarily to drains and submarine discharge, but also flow to the Sparling 2 well. The flow paths originating farther west have long travel times to drains or the sea—50 to more than 100 years (using porosities of 0.4 for layers 1 and 5, 0.3 for layers 2 and 4, and 0.2 for layer 3, from Fetter, 1994). Particles originating farther east have progressively shorter travel times—years to decades. These long travel times also suggest the effects of pumping water out of the aquifer near the ends of the particle tracks would take a similarly long time for the response to that pumpage to reach steady state (where an aquifer is unconfined).

Another important aspect of the forward tracking results for the entire model area is that most of particles discharge to submarine discharge or springs instead of to Chimacum Creek. This indicates the flow system is highly influenced by these boundary conditions—much more so than the boundary cells that represent the stream. This situation prevents the calculation of percentage of pumpage captured from the stream over some period of time for various locations or depths in transient simulations. Leake and others (2008, p. 7) states: "If a model includes head-dependent flow boundaries that are not physical hydrologic features, for example specified-head cells at the lateral boundary of a model, the mapped capture values can be erroneous where withdrawals by wells induce inflow from or reduce outflow to those boundaries." Except for the southwestern part of the model, particle flow paths are dominated by such non-physical hydrologic features (fig. 16).

Particle-tracking with the model shows that the source of water to the Sparling 2 well, which withdraws water from unit LA (model layer 4) primarily is from the till plain to the southwest (fig. 17), with only a few particles backtracking to Chimacum Creek. This does not imply that the Sparling 2 well does not intercept water that may have eventually discharged to the creek.

Forward- and reverse-tracking of particles from and to Chimacum Creek shows that water flowing into the creek (blue lines in fig. 18) is primarily in the headwaters and near the mouth, whereas water flowing from the creek into the groundwater system (red lines) in the western part of the model eventually discharge back into the creek, and in the eastern part flow to seeps and springs along the coast and submarine discharge to Puget Sound.

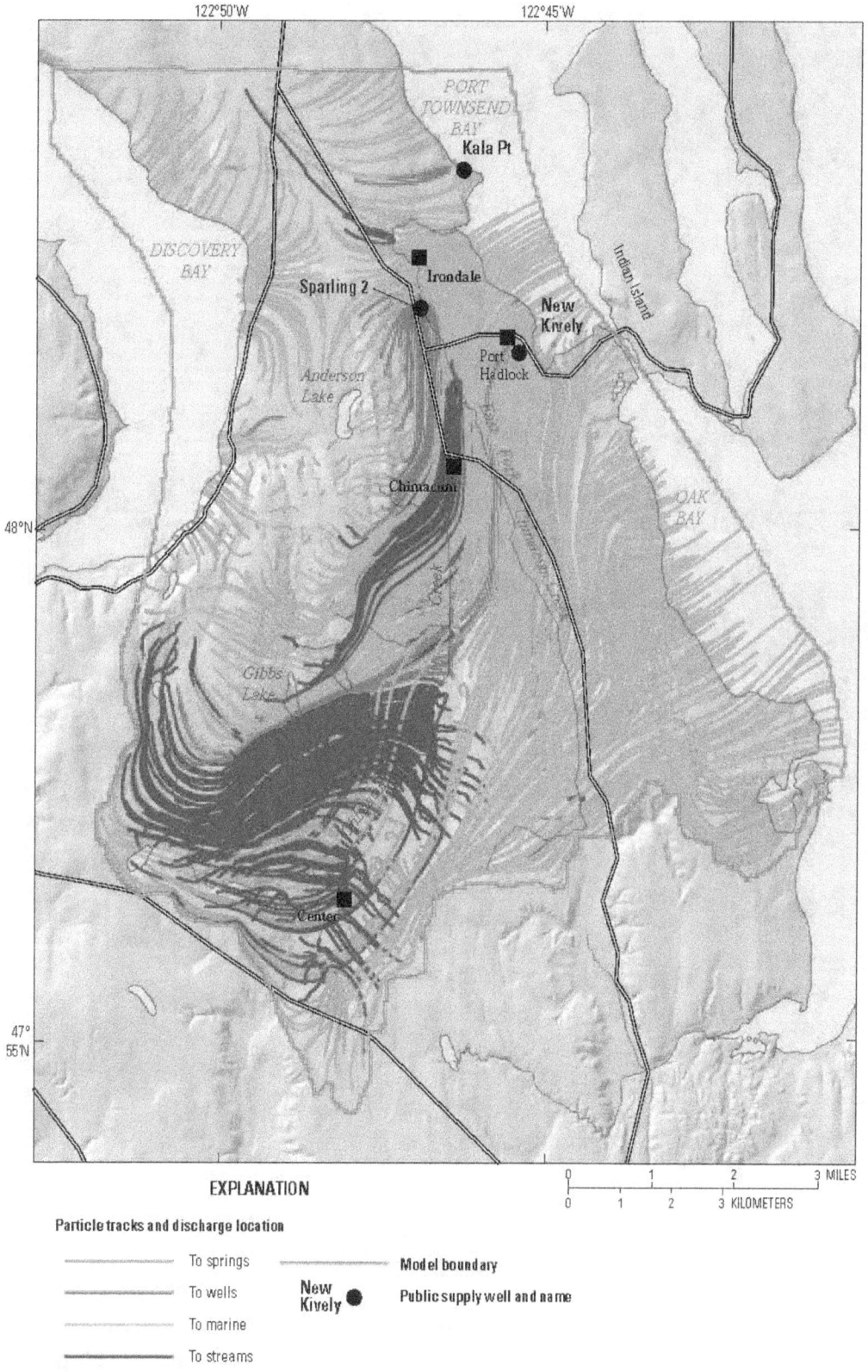

Figure 16. Forward particle tracking from topmost layer to sinks, colored by boundary condition at particle terminus, Chimacum Creek Basin and vicinity, Jefferson County, Washington.

EXPLANATION

———— Till

———— Creek

● Well

Figure 17.   Reverse particle tracking from the Jefferson County Public Utility District Number 1 Sparling well, Chimacum Creek Basin and vicinity, Jefferson County, Washington.

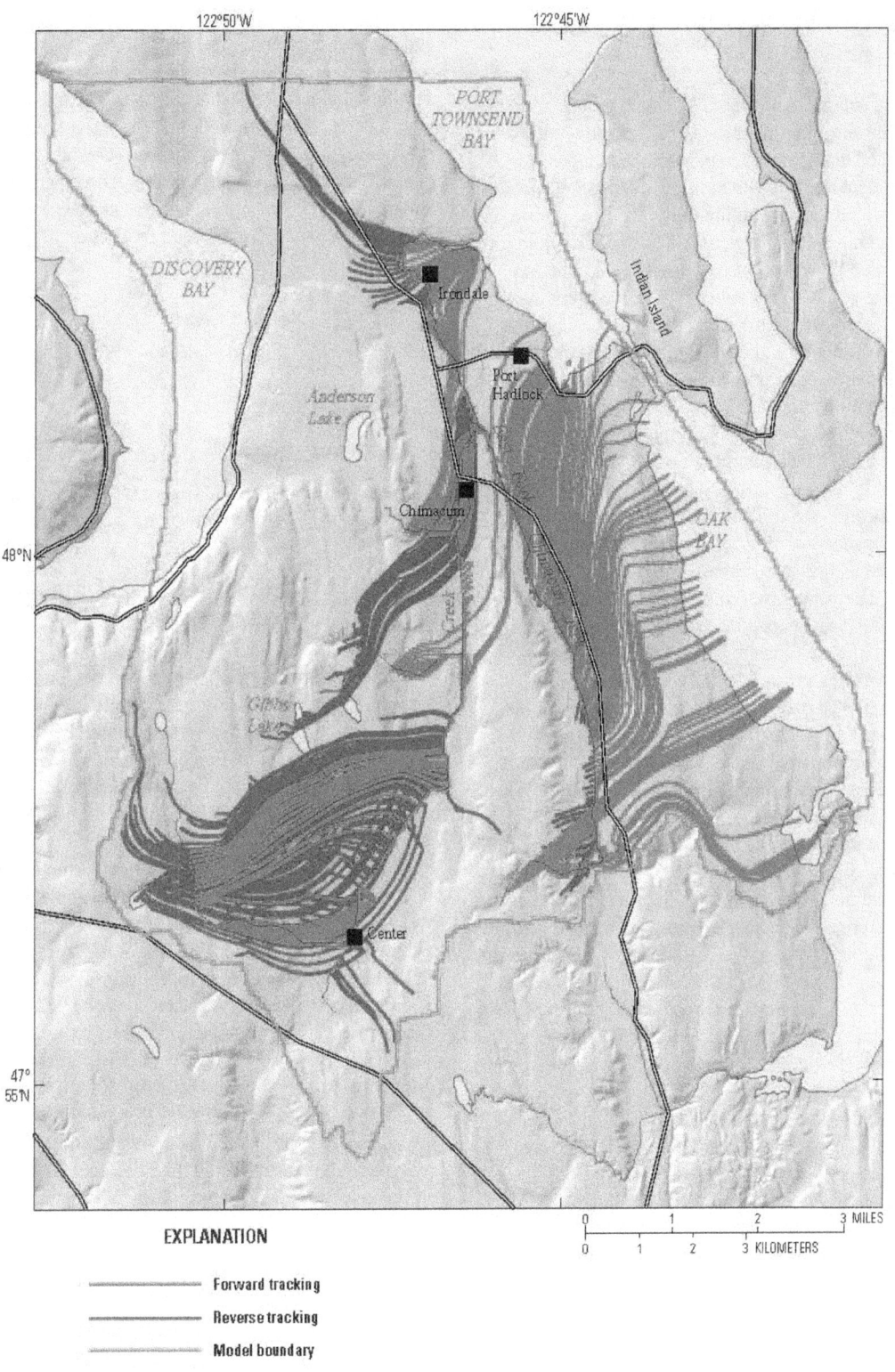

Figure 18.   Forward and reverse particle tracking to and from Chimacum Creek and East Fork Chimacum Creek, Jefferson County, Washington.

Areal Variation of Response Coefficient for a Well

For a transient groundwater flow model with few or no boundary conditions other than a stream, "capture maps" may be produced that show what percentage of a well discharge was captured from streamflow over some period of time. Although transient capture maps for some period of time (for example, 10 or 30 years) are not appropriate for systems dominated by non-physical boundary conditions (Leake and others, 2008), the equivalent of a capture map for steady-state conditions may be produced by calculating the response coefficient for a well. The response coefficient for a well in a steady-state simulation is a measure of the change in streamflow due to an introduced pumping well as a percentage of the well discharge. Figure 19 is a capture map produced by adding a well pumping 68,583 ft^3/d (based on the average pumping rate at PUD #1 Sparling 2 well) at various hypothetical locations in the Lower Aquifer unit (LA, model layer 4) to the steady-state Current Conditions simulation. The change in flow in Chimacum Creek with the additional well pumpage was determined from the simulated groundwater budget and plotted as a percentage of the withdrawal rate as a function of the well location. To consider various well locations within the aquifer, 406 simulations were run.

In the southwest part of the model, the capture rates are low near Peterson Lake, as the steady state model predicted well pumpage would largely take water from the lake. It should be noted, however, that the steady state model does not consider the possibility that the lake could eventually go dry. In the southeastern part of the model, capture rates decline as the simulated well becomes farther from the creek and closer to constant head boundary conditions. In the northern part of the model, capture rates decrease rapidly as the well begins extracting water that would otherwise discharge to Discovery or Port Townsend Bays.

A steady state model can be considered a transient model of an infinite time period, and capture rates for shorter periods would be lower. The term "capture" includes extracting water, not from the creek directly, but also intercepting water that may have flowed to the creek or influenced the direction of other water that would have otherwise flowed to the creek.

These "capture" rates correspond to the response coefficients that Groundwater Management software (GWM, Ahlfeld and others, 2005) calculates to estimate optimal pumping times and rates for transient simulations. For steady state simulations, the timing is excluded from consideration, and optimal pumping rates are based exclusively on the response coefficient. Thus, GWM results for a transient simulation would show that the optimal pumping scenario would be pumping the well with the lowest response coefficient to its maximum allowed (or possible) rate, then pumping the well with the next lowest response coefficient, and so forth. The response coefficients calculated by GWM for five PUD #1 wells, from lowest to highest, are:

1. Olympic Mobile Village: 28
2. Seton (Willison): 33
3. Airport 2B: 64
4. New Kively: 82
5. Sparling 2: 96

These response coefficients agree with the percentage capture rates in figure 19 at the corresponding well locations.

Effect of Irrigation Wells and Depth on Chimacum Creek Streamflow

The calibrated steady-state Current Conditions simulation was used to estimate the effect of irrigation wells by (1) removing the irrigation wells and their return flow at the surface and (2) moving the irrigation wells to a shallower aquifer. The withdrawal rate for all the irrigation wells combined was about 0.45 ft^3/s, or a total of 329 acre-ft/yr. Streamflow decreased by 0.39 ft^3/s when the irrigation wells were removed because the return flows from the irrigation wells no longer provided recharge. As simulated, irrigation wells transfer water from the deep aquifers to the land surface where the return flow recharges groundwater that discharges to the stream. Locating the irrigation wells in shallow aquifers slightly decreased streamflow by 0.08 ft^3/s because shallow irrigation wells capture a part of the water that would otherwise discharge to the stream. These streamflow values are relatively small and would be difficult to measure. A USGS streamflow measurement of 7.5 ft^3/s, which is the steady-state streamflow condition at WDOE gaging station 17B050, rated as "good" indicates that the measurement is within 5 percent of the true value (Rantz and others, 1982), or ±0.375 ft^3/s. The streamflow differences determined in this simulation are within, or just outside, those error bands.

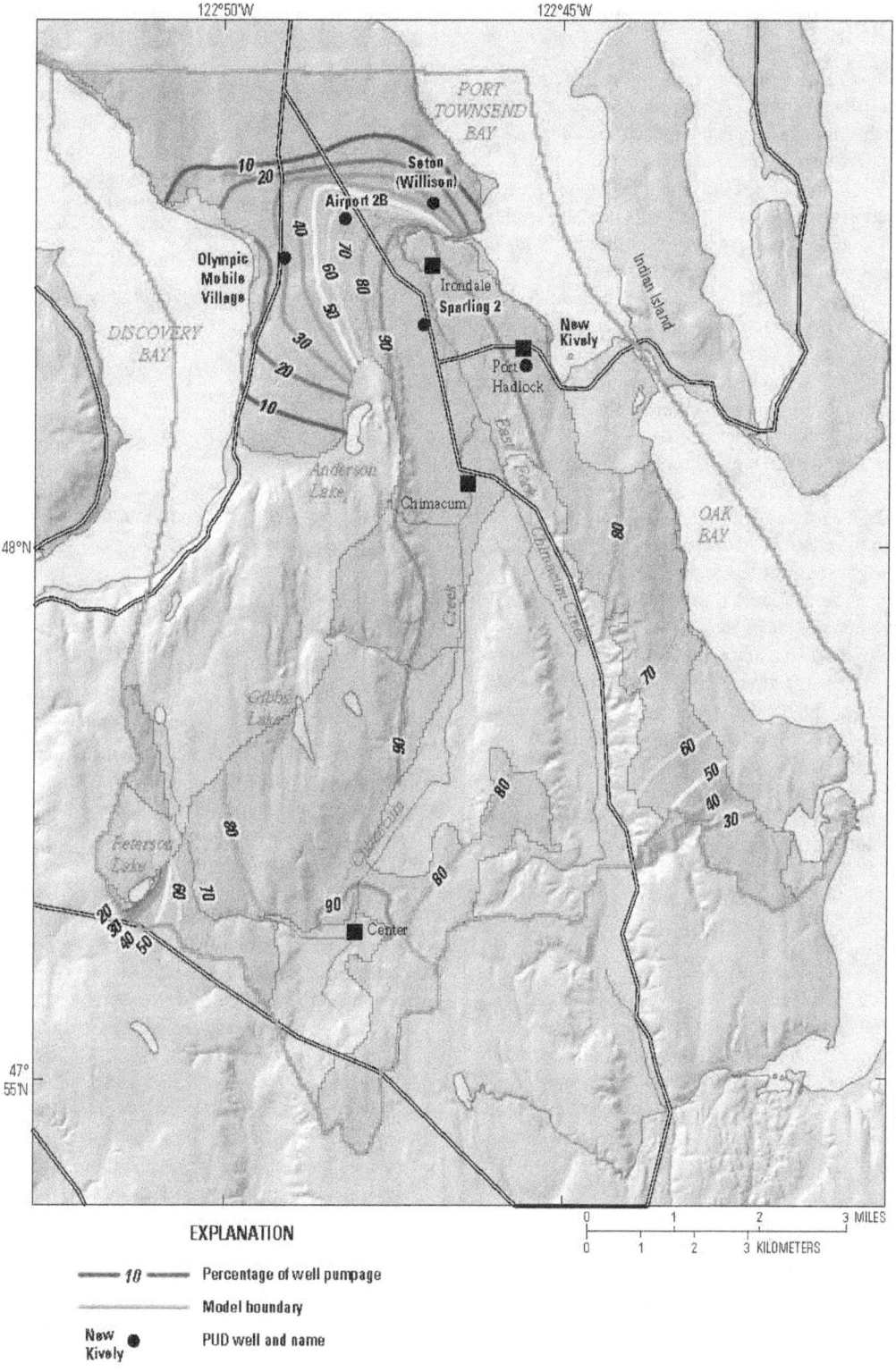

Figure 19. Response coefficient (percentage of well pumpage captured from streamflow) for a well pumping 68,583 cubic feet per day at any location in model layer 4 (Lower Aquifer), Chimacum Creek Basin and vicinity, Jefferson County, Washington.

Suggestions for Further Study

Simulations of transient conditions might allow some insight into the effects of pumping stresses that occur irregularly. Such simulations, however, could prove intractable if the groundwater-surface water interactions are highly variable in time and space, or if the quantities involved are very small fractions of the overall mass balance. A possible solution to this would be to apply newer solution methods better at handling numerical instabilities, such as provided by MODFLOW-NWT, which could provide successful convergence within transient condition timesteps. It would also allow for simulation of conceptually unconfined model layers as variable-saturated-thickness layers, which would alleviate concerns about the adequacy of simulating them as constant-saturated-thickness layers.

The effect of the extensive organic deposits in the East Fork Chimacum Creek mimic bank storage effects seen in surface-water hydrographs after high flows; however, the time scale is on the order of months rather than days. Such delayed release of groundwater storage to surface water may be more common in other areas than is typically recognized, due to the lower magnitude of such effects that likely occur in less organic or extensive deposits. An empirical study of the storage and release of groundwater in this deposit would help characterize its interaction with the groundwater and surface-water systems, and possibly lead to a method of treating it in a numerical simulation. Such a study would also benefit any future assessment of water quality in Chimacum Creek, due to the high levels of organic constituents in the water discharging from those deposits.

Summary

Growth of a population dependent exclusively on groundwater for water supplies, and widespread concerns about adequate base flow for fish listed in accordance with the Endangered Species Act of 1973, has led to a need for better understanding of the groundwater system in northeast Jefferson County and its interaction with Chimacum Creek. A groundwater-flow model was developed to evaluate potential future effects of growth and of water-management strategies on water resources in the Chimacum Creek Basin. The model covers an area of about 64 square miles (mi^2) on the Olympic Peninsula in northeastern Jefferson County, Washington. The Chimacum Creek Basin drains an area of about 53 mi^2 and consists of Chimacum Creek and its tributary East Fork Chimacum Creek, which converge near the town of Chimacum and discharge to Port Townsend Bay near the town of Irondale. The topography of the model area consists of north-south oriented, narrow, regularly spaced parallel ridges and valleys that are characteristic of a fluted glaciated surface. Thick accumulations of peat occur along the axis of larger valleys and provide rich soils for agricultural use. The study area is underlain by a north-thickening sequence of unconsolidated glacial (till and outwash) and interglacial (fluvial and lacustrine) deposits, and sedimentary and igneous bedrock units that crop out along the margins and the western interior of the model area. Six hydrogeologic units form the basis of the groundwater-flow model:

1. Upper Confining unit (UC, roughly corresponding to Quaternary Alluvium

2. Upper Aquifer unit (UA, roughly corresponding to Vashon Recessional Outwash)

3. Middle Confining unit (MC, roughly corresponding to Vashon Till)

4. Lower Aquifer unit (LA, roughly corresponding to Vashon Advance Outwash)

5. Lower Confining unit (LC, undifferentiated pre-Vashon unconsolidated deposits)

6. Bedrock.

Groundwater flow in the Chimacum Creek Basin and vicinity was simulated using the groundwater-flow model, MODFLOW-2005. The finite-difference model grid comprises 245 columns, 313 rows, and 6 layers; each model cell has a horizontal dimension of 200 × 200 feet (ft). The thickness of model layers varies throughout the model area. Boundary conditions representing inflow and outflow components were implemented using packages in MODFLOW-2005. The Recharge Package was used to represent recharge from precipitation and water returned to the groundwater system through seepage from septic systems, deep percolation of irrigation water, and public water-system conveyance losses. The WELL Package was used to represent withdrawals from wells and return flows associated with withdrawals from domestic wells. Streams, springs, and groundwater seeps were simulated with either the Stream or the Drain packages, and the General-Head Boundary Package was used to represent the exchange of water along the northern model boundary.

Groundwater flow was calibrated to steady-state conditions using average recharge, discharge, and water levels for the 180-month period October 1994–September 2009. During model calibration, variables were adjusted within probable ranges to minimize differences between measured and simulated groundwater levels and stream base flows. The model as calibrated to steady-state conditions has a mean residual of 4.5 ft and a standard error on the mean of 2.1 ft for heads and, for flows, 0.64 ± 0.42 cubic feet per second.

Simulated steady-state inflow to the model area from precipitation and secondary recharge, or "return flow," was 16,347 acre-feet per year (acre-ft/yr); groundwater inflow from other basins to the north of the model boundary was 1,518 acre-ft/yr (net, 3,114 acre-ft/yr in and 1,596 acre-ft/yr out) and simulated inflow from lake leakage was 613 acre-ft/yr (net, 684 acre-ft/yr in and 71 acre-ft/yr out). Simulated outflow from the model primarily was through discharge to Puget Sound (10,022 acre-ft/yr), streams (5,424 acre-ft/yr), springs and seeps (1,521 acre-ft/yr), and withdrawals from wells (1,506 acre-ft/yr).

Four simulations were formulated using the calibrated model; one to represent current conditions (2009, the end of the period used for calibration); and three to provide representative examples of how the model can be used to evaluate the relative effects of potential changes in groundwater withdrawals and consumptive use on groundwater levels and stream base flows: Probable Future Use, based on population projections; Full Beneficial Use, based on Jefferson County Public Utility District #1 water rights; Sanitary Sewer, based on eliminating septic return flows in the Urban Growth Area. Particle tracking was used to assess flowpaths from sources and to sinks, and the effects of the presence of irrigation wells and their depths was assessed.

Further study may include transient simulations to measure the effect of stresses on the groundwater system due to seasonal pumping. Study of the delayed bank storage effect of the extensive organic deposits in the East Fork Chimacum Creek may provide insight as to whether this is a significant factor in other basins in the Puget Lowlands.

Acknowledgments

Thanks to Bill Graham, Resource Manager, Jefferson County Public Utility District #1 for collecting monthly water-level measurements, to Tom Culhane and Dave Nazy, Washington State Department of Ecology, and to Neil Harrington and Mike Dawson, Jefferson County Environmental Health Specialists.

References Cited

Ahlfeld, D.P., Barlow, P.M., and Mulligan, A.E., 2005, GWM—A ground-water management process for the U.S. Geological Survey Modular Ground-Water Model (MODFLOW-2000): U.S. Geological Survey Open-File Report 2005-1072, 124 p.

Bear, Jacob, 1979, Hydraulics of groundwater: New York, McGraw-Hill, 569 p.

Bidlake, W.R., and Payne, K.L., 2001, Estimating recharge to ground water from precipitation at Naval Submarine Base, Bangor and vicinity, Kitsap County, Washington: U.S. Geological Survey Water-Resources Investigations Report 01-4110, 33 p.

de Marsily, Ghislain, Lavedan, G., Boucher, M., and Fasanino, G., 1984, Interpretation of interference tests in a well field using geostatistical techniques to fit the permeability distribution in a reservoir model, in Verly, George, and others, eds., Geostatistics for natural resources characterization, Part 2: Dordrecht, Netherlands, D. Reidel Publishing Company, p. 831–849.

Doherty, J., 2003, Ground water model calibration using pilot points and regularization: Ground Water, v. 41, no. 2, p. 170–177.

Doherty, J., 2005, PEST—Model-independent parameter estimation: Corinda, Australia, Watermark Numerical Computing, variously paged.

Doherty, J., and Hunt R.J., 2009, Two statistics for evaluating parameter identifiability and error reduction: Journal of Hydrology, v. 366, no. 1–4, p. 119–127

Doherty, J.E., and Hunt, R.J., 2010, Approaches to highly parameterized inversion—A guide to using PEST for groundwater-model calibration: U.S. Geological Survey Scientific Investigations Report 2010–5169, 59 p.

Easterbrook, D.J., 1979, The last glaciation of northwest Washington, in Armentrout, J.M., and others, eds., Cenozoic paleogeography of the western United States, Pacific Coast Paleogeography Symposium, No. 3: Society of Economic Paleontologists and Mineralogists Symposium, , p. 177–189.

Faunt, C.C., Provost, A.M., Hill, M.C., and Belcher, W.R., 2011, Comment on "An unconfined groundwater model of the Death Valley Regional Flow System and a comparison to its confined predecessor" by R.W.H. Carroll, G.M. Pohll and R.L. Hershey [Journal of Hydrology 373/3-4, pp. 316-328]: Journal of Hydrology, v. 397, no. 3-4, p. 306-309. (Also available at http://digitalcommons.unl.edu/usgsstaffpub/394.)

Fetter, C.W., 1994, Applied Hydrogeology (3rd ed.): Upper Saddle River, New Jersey, Prentice Hall, 691 p.

Freeze, R.A., and Cherry, J.A., 1979, Ground water: Englewood Cliffs, New Jersey, Prentice-Hall, 604 p.

Golder Associates, 2008, Groundwater storage in the Chimacum Creek Basin, Draft step report, Screening of prospective sources and recharge sites: Prepared for Public Utility District #1 of Jefferson County, Washington and the Water Resources Inventory Area 17 Planning Unit, Golder Associates, 32 p.

Golder Associates, 2010, Water demand and availability assessment for WRIA 17–Stage 1: Prepared for the Water Resources Inventory Area (WRIA) 17 Planning Unit, Golder Associates, 44 p.

Grimstad, Peder, and Carson, R.J., 1981, Geology and groundwater resources of eastern Jefferson County, Washington: Washington State Department of Ecology Water Supply Bulletin No. 54, 125 p.

Harbaugh, A.W., 2005, MODFLOW-2005, the U.S. Geological Survey modular ground-water model—The ground-water flow process: U.S. Geological Survey Techniques and Methods 6-A16, variously paged. (Also available at http://water.usgs.gov/nrp/gwsoftware/modflow.html.)

Hoekstra, A.Y., Chapagain, A.K., Aldaya, M.M. and Mekonnen, M.M., 2011, The water footprint assessment manual: Setting the global standard, Earthscan, London, UK

Jones, J.L., Welch, W.B., Frans, L.M., and Olsen, T.D., 2011, Hydrogeologic framework, groundwater movement, and water budget in the Chimacum Creek basin and vicinity, Jefferson County, Washington: U.S. Geological Survey Scientific Investigations Report 2011–5129, 28 p. (Also available at http://pubs.usgs.gov/sir/2011/5129/.)

Kuniansky, E.L., Gómez-Gómez, Fernando, and Torres-González, Sigfredo, 2004, Effects of aquifer development and changes in irrigation practices on ground-water availability in the Santa Isabel area, Puerto Rico: U.S. Geological Survey Water-Resources Investigations Report 03-4303, 65 p. (Also available at http://pubs.usgs.gov/wri/wri034303/.)

LaVenue, A.M., and Pickens, J.F., 1992, Application of a coupled adjoint sensitivity and kriging approach to calibrate a groundwater flow model: Water Resources Research, v. 28, no. 6, p. 1,543–1,569.

Leake, S.A., Pool, D.R., and Leenhouts, J.M., 2008, Simulated effects of ground-water withdrawals and artificial recharge on discharge to streams, springs, and riparian vegetation in the Sierra Vista subwatershed of the Upper San Pedro Basin, southeastern Arizona: U.S. Geological Survey Scientific Investigations Report 2008-5207, 14 p. (Also available at http://pubs.usgs.gov/sir/2008/5207/.)

McDonald, M.G., and Harbaugh, A.W., 1988, A modular three-dimensional finite-difference ground-water flow model: U.S. Geological Survey Techniques of Water-Resources Investigations, Book 6, Chapter Al, 586 p. (Also available at http://pubs.usgs.gov/twri/twri6al/.)

National Oceanic and Atmospheric Administration, 2006, Endangered Species Act status of west coast Salmon and Steelhead: National Oceanic and Atmospheric Administration, accessed July 28, 2011, at http://www.oregon.gov/OWEB/docs/board/2006-01/itemf_attc.pdf.

National Oceanic and Atmospheric Administration, 2007, Climatological data—Annual summary, Washington: Asheville, North Carolina, National Climatic Data Center, v. 111, no. 13, 30 p.

National Oceanic and Atmospheric Administration, NOAA's 1981-2010 Climate Normals: National Climatic Data Center, 2011, accessed June 10, 2013, at http://www.ncdc.noaa.gov/oa/climate/normals/usnormals.html NOAA's 1981-2010 Climate Normals.

Parametrix, Inc., Pacific Groundwater Group, Inc., Montgomery Water Group, Inc., and Caldwell and Associates, 2000, Stage 1 technical assessment as of February 2000, Water Resource Inventory Area (WRIA) 17: Jefferson County, Washington, Project No. 553-1820-007, accessed July 28, 2011, at http://www.ecy.wa.gov/programs/eap/wrias/Planning/docs/063010_wria17_water_supply_demand.pdf.

Petkewich, M.D., and Campbell, B.G., 2007, Hydrogeology and simulation of ground-water flow near Mount Pleasant, South Carolina—Predevelopment, 2004, and predicted scenarios for 2030: U.S. Geological Survey Scientific Investigations Report 2007-5126, 79 p. (Also available at http://pubs.usgs.gov/sir/2007/5126/.)

Pollock, D.W., 1994, User's guide for MODPATH/MODPATH-PLOT, Version 3—A particle tracking post-processing package for MODFLOW, the U.S. Geological Survey finite-difference ground-water flow model: U.S. Geological Survey Open-File Report 94-464, 242 p. (Also available at http://pubs.er.usgs.gov/publication/ofr94464.)

Prudic, D.E., 1989, Documentation of a computer program to simulate stream-aquifer relations using a modular, finite-difference, ground-water flow model: U.S. Geological Survey Open-File Report 88-729, 113 p. (Also available at http://pubs.er.usgs.gov/publication/ofr88729.)

Rantz, S.E., and others, 1982, Measurement and computation of streamflow: U.S. Geological Survey Water-Supply Paper 2175. (Also available at http://pubs.usgs.gov/wsp/wsp2175/.)

Ritter, D.F., 1978, Process geomorphology: Dubuque, Iowa, William C. Brown Company Publishers, 414 p.

Simonds, F.W., Longpré, C.I., and Justin, G.B., 2004, Groundwater system in the Chimacum Creek Basin and surface-water/groundwater interaction in Chimacum and Tarboo Creeks and the Big and Little Quilcene Rivers, Eastern Jefferson County, Washington: U.S. Geological Survey Scientific Investigations Report 2004–5058, 49 p. (Also available at http://pubs.er.usgs.gov/publication/sir20045058.)

Sloto, R.A., and Crouse, M.Y., 1996, HYSEP—A computer program for streamflow hydrograph separation and analysis: U.S. Geological Survey Water-Resources Investigations Report 96-4040, 46 p. (Also available at http://pubs.er.usgs.gov/publication/wri964040.)

Tabor, R.W., and Cady, W.M., 1978, Geologic map of the Olympic Peninsula, Washington: U.S. Geological Survey Miscellaneous Investigation Series Map I-994, scale 1:125,000.

Troutman, B.M., 1985, Errors and parameter estimation in precipitation-runoff modeling 2—Case study: Water Resources Research, v. 21, no. 8, p. 1,214–1,222.

Washington State Department of Ecology, 2003, Well Logs, accessed June 10, 2013 at http://apps.ecy.wa.gov/welllog/.

www.ingramcontent.com/pod-product-compliance
Lightning Source LLC
Chambersburg PA
CBHW081554170526
45166CB00009B/2694